CHOOSING AND USING STATISTICS
A BIOLOGIST'S GUIDE

Choosing and Using Statistics

A Biologist's Guide

Calvin Dytham

Department of Biology, University of York

Second Edition

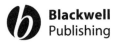
Blackwell
Publishing

BLACKWELL PUBLISHING
350 Main Street, Malden, MA 02148-5020, USA
108 Cowley Road, Oxford OX4 1JF, UK
550 Swanston Street, Carlton, Victoria 3053, Australia

First published 1999
Second edition 2003
Reprinted 2004, 2005

Library of Congress Cataloging-in-Publication Data

Dytham, Calvin.
    Choosing and using statistics: a biologist's guide/Calvin Dytham.—2nd ed.
        p. cm.
    Includes index.
    ISBN 1-4051-0243-8 (pbk.: alk. paper)
    1. Biometry. I. Title.

    QH323.5.D98 2003
    570′.1′5195—dc21

                                                                2002074760

A catalogue record for this title is available from the British Library.

Set in 9½/12pt Berling
by Graphicraft Ltd, Hong Kong
Printed and bound in the United Kingdom
by TJ International Ltd, Padstow, Cornwall

The publisher's policy is to use permanent paper from mills that operate a sustainable
forestry policy, and which has been manufactured from pulp processed using acid-free and
elementary chlorine-free practices. Furthermore, the publisher ensures that the text paper
and cover board used have met acceptable environmental accreditation standards.

For further information on
Blackwell Publishing, visit our website:
www.blackwellpublishing.com

Contents

8 The tests 2: tests to look at relationships, 165

9 The tests 3: tests for data exploration, 200

Preface

My aim was to produce a statistics book with two characteristics: to assume that the reader is using a computer to analyse data and to contain absolutely no equations.

This book is a handbook for biologists who want to process their data through a statistical package on the computer, to select the most appropriate methods and extract the important information from the, often confusing, output that is produced. It is aimed, primarily, at undergraduates and masters students in the biological sciences who have to use statistics in practical classes and projects. Such users of statistics do not have to understand exactly how the test works or how to do the actual calculations and these things are not covered in this book as there are more than enough books that do that already. What is important is that the right statistical test is used and the right inferences made from the output of the test. An extensive key to statistical tests is included for the former and the bulk of the book is made up of descriptions of how to carry out the tests to address the latter.

In several years of teaching statistics to biology students it is clear to me that most students do not really care how or why the test works. They do care a great deal that they are using an appropriate test and interpreting the results properly. I think that this is a fair aim to have for occasional users of statistics. Of course, anyone going on to use statistics frequently should become familiar with the way that calculations manipulate the data to produce the output as this will give a better understanding of the test.

If this book has a message it is this: *think about the statistics **before** you collect the data!* So many times I have seen rather distraught students unable to analyse their precious data because the experimental design they used was inappropriate. On such occasions I try to find a compromise test that will make the best of a bad job but this often leads to a weaker conclusion than might have been possible if more forethought had been applied from the outset. There is no doubt that if experiments or sampling strategies are designed with the statistics in mind better science will result.

Statistics are often seen by students as the 'thing you must do to data at the end'. Please try to avoid falling into this trap yourself. 'Thought' experiments producing dummy data are a good way to try out experimental designs and are much less labour intensive than real ones!

Although there are no equations in this book I am afraid there was no way to totally avoid statistical jargon. To ease the pain somewhat, an extensive glossary and key to symbols is included. So when you are navigating your way through the key to choosing a test you should look up any words you do not understand.

In this book I have given extensive instructions for the use of three widely available software packages: SPSS, MINITAB and Excel. However, the key to choosing a statistical test is not at all package specific so if you use a software package other than the three I focus on, or if you a using a calculator, you will still be able to get a good deal out of this book.

If every sample gave the same result there would be no need for statistics. All aspects of biology seem to be filled with variation. It is statistics that can be used to penetrate the haze of experimental error and the inherent variability of the natural world to reach the underlying causes and processes at work. So, try not to hate statistics, they are merely a tool that, when used wisely and properly, can make the life of a biologist much simpler and give conclusions a sound basis.

The second edition

In the 4 years since I wrote the first edition of this book there have been several new versions of the software produced. This was inevitable, and has been the main driver for the production of a second edition using more recent versions of the software. There are few changes to the structure and message of the book from the first edition. I have removed the mini-reviews of other statistics books and feel that anyone moving on from this book should consider moving to either Zar's *Biostatistical Analysis* or Sokal and Rohlf's *Biometry*. Both are large, explain how to calculate statistical tests and are more comprehensive than this volume.

I have received many comments about the first edition and I am grateful for the many suggestions on how to improve the text and coverage. Requests to add further statistical packages have been the most common suggestion for change. But, as there was surprisingly little consensus on the packages to add, I decided to restrict the coverage, as before, to SPSS, MINITAB and Excel. I apologize to all those users of Systat, Genstat, SAS, Statistica, S/S-plus/R and GLIM who were hoping their package would be included.

I have expanded some sections from the first edition. This includes a simple table to encapsulate the key, coverage of the G-test, more ANOVA designs, logistic regression and some other regression techniques. There are also more examples of multivariate techniques. I have tried to do this without shifting the audience and suggest that anyone needing to use a range of multivariate techniques should look elsewhere for guidance.

How to use this book

This is definitely not a book that should be read from cover to cover. It is a book to refer to when you need assistance with statistical analysis, either when choosing an appropriate test or when carrying it out. The basics of statistical analysis and experimental design are covered briefly, but these sections are intended mostly as a revision aid, or outline of some of the more important concepts. Other statistics books may help you choose those that are most appropriate for you if you want or need more details.

The heart of the book is the key in Chapter 3. The rest of the book hinges on the key, explaining how to carry out the tests, giving assistance with the statistical terms in the glossary or giving tips on the use of computers and packages.

Packages used

MINITAB® version 13.10 and 13.32, *MINITAB inc.*
SPSS® version 10.1, *SPSS inc.*
Excel™ versions 97 and 2000, *Microsoft Corporation.*
Running on: Windows® versions 95 (release 2), 98, ME and 2000, *Microsoft Corporation.*

Example data

In the spirit of dummy data collection, all example data used throughout this book have been fabricated. Any similarity to data alive or dead is purely coincidental.

Acknowledgements for the first edition

Thanks to Sheena McNamee for support during the writing process, to Andrea Gillmeister and two anonymous reviewers for commenting on an early version of the manuscript and to Terry Crawford, Jo Dunn, David Murrell and Josephine Pithon for recommending and lending various books. Thanks also to Ian Sherman and Susan Sternberg at Blackwells and to many of my colleagues who told me that the general idea of a book like this was a sound one. Finally, I would especially like to thank the students at the University of York, UK, who brought me the problems that provided the inspiration for this book.

Acknowledgements for the second edition

Thanks to all the many people who contacted me with suggestions and comments about the first edition. I hope you can see that many of the corrections and improvements have come directly from you. Five anonymous reviewers provided many useful comments about the proposal for a second edition. Thanks to Sarah Shannon, Cee Brandson, Katrina McCallum and many others at Blackwells for seeing this book through and especially for producing a second superb and striking cover. S'Albufera Natural Parc and Nick Riddiford provided a very convenient bolt-hole for writing. Once again, I give special thanks to Sheena and to my colleagues, PhD students and undergraduate students at the University of York. Finally, thanks to everyone on the MRes EEM course over the last 4 years.

Calvin Dytham
York, 1998 and 2002

Eight steps to successful data analysis

This is a very simple sequence that, it you follow it, will integrate the statistics you use into the process of scientific investigation. As I make clear here, statistical tests should be considered *very early* in the process and not left until the end.

1 Decide what you are interested in.

2 Formulate a hypothesis or several hypotheses (see Chapters 2 and 3 for guidance).

3 Design the experiment, manipulation or sampling routine that will allow you to test the hypotheses (see Chapters 2 and 4 for some hints on how to go about this).

4 *Collect dummy data* (i.e. make up approximate values based on what you expect to obtain). The collection of 'dummy data' may seem strange but it will convert the proposed experimental design or sampling routine into something more tangible. The process can often expose flaws or weaknesses in the data collection routine that will save a huge amount of time and effort.

5 Use the key (presented in Chapter 3) to guide you towards the appropriate test or tests.

6 Carry out the test(s) using the dummy data. (Chapters 6–9 will show you how to input the data, use the statistical packages and interpret the output.)

7 If there are problems go back to step 3 (or 2), otherwise proceed to the collection of the real data.

8 Carry out the test(s) using the real data. Report the findings and/or return to step 2.

I implore you to use this sequence. I have seen countless students who have spent a long time and a lot of effort collecting data only to find that the experimental or sampling design was not quite right. The test they are forced to use is much less powerful than one they could have used with only a slight change in the experimental design. This sort of experience tends to turn people away from statistics and become 'scared' of them. This is a great shame as statistics are a hugely useful and vital tool in science.

The rest of the book follows this eight step process but you should use it for guidance and advice when you become unsure of what to do.

2

The aim of this section is to introduce, in rather broad terms, some of the recurring concepts of data collection and analysis. Everything introduced here is covered at greater length in later chapters and certainly in the many statistics text books that aim to introduce statistical theory and experimental design to scientists.

The key to statistical tests in the next chapter assumes that you are familiar with most of the basic concepts introduced here.

Observations

These are the raw material of statistics and can include anything recorded as part of an investigation. They can be on any scale from a simple 'raining or not raining' dichotomy to a very sophisticated and precise analysis of nutrient concentrations. The type of observations recorded will have a great bearing on the type of statistical tests that are appropriate.

Observations can be simply divided into three types: *categorical* where the observations can be in a limited number of categories that have no obvious scale (e.g. 'oak', 'ash', 'elm'); *discrete* where there is a real scale but not all values are possible (e.g. 'number of eggs in a nest' or 'number of species in a sample'); and *continuous* where any value is theoretically possible but is restricted by the measuring device (e.g. lengths, concentrations).

Different types of observations are considered in more detail in Chapter 5.

Hypothesis testing

The cornerstone of scientific analysis is hypothesis testing. The concept is rather simple: almost every time a statistical test is carried out it is testing the probability that a hypothesis is correct. If the probability is small then the hypothesis is deemed to be untrue and it is rejected in favour of an alternative. This is done in what seems to be a rather upside down way as the test is always of what is called the null hypothesis rather than the more interesting hypothesis. The null hypothesis is the hypothesis that nothing is going on (it is often labelled as H_0). For example, if the weights of bulbs for two cultivars of daffodils were being investigated, the null hypothesis would be that there is nothing going on: 'the weights of the two groups of bulbs are the same'. A statistical test is carried out to find out how likely that null hypothesis is to be true. If we decide to reject the null hypothesis we must accept the alternative, more interesting hypothesis (H_1) that: 'the weights of bulbs for the two cultivars are different'.

P-values

The *P*-value is the bottom line of most statistical tests. It is simply the probability that the hypothesis being tested is true. So if a *P*-value is given as 0.06, that indicates that the hypothesis has a 6% chance of being true. In biology it is usual to take a value of 0.05 or 5% as the critical level for the rejection of a hypothesis. This means that providing a hypothesis has a less than one in 20 chance of being true, we reject it. As it is the null hypothesis that is nearly always being tested we are always looking for low *P*-values to reject this hypothesis and accept the more interesting alternative hypothesis.

Clearly the smaller the *P*-value the more confident we can be in the conclusions drawn from it. A *P*-value of 0.0001 indicates that the chance of the hypothesis being tested being true is one in 10 000 and this is much more convincing than a marginal $P = 0.049$.

P-values and the types of errors that are implicitly accepted by their use are considered further in Chapter 4.

Sampling

Observations have to be collected in some way. This process of data acquisition is called sampling. Although there are almost as many different methods that can be used for sampling as there are possible things to sample, there are some general rules. One of the most obvious is that more observations are usually better than few. Balanced sampling is also important (i.e. when comparing two groups take the same number of observations from each group).

Most statistical tests assume that samples are taken at random. This sounds easy but is actually quite difficult to achieve. For example, if you are sampling beetles from pitfall traps the sample may seem totally random but in fact is quite biased towards those species that move around the most. Another common bias is to chose a point at random and then measure the nearest individual to that point assuming that this will produce a random sample. It will not be random at all as individuals at the edges of clumps are more likely to be selected than those in the middle. There are methods available to reduce problems associated with non-random sampling but the first step is to be aware of the problem.

A further assumption of sampling is that individuals are either only measured once or they are all sampled on several occasions. This assumption is often violated if, for example, the same site is visited on two occasions and the same individuals or clones are inadvertently remeasured.

The sets of observations collected are called variables. A variable can be almost anything it is possible to record as long as different individuals can be assigned different values.

Some of the problems of sampling are considered in Chapter 4.

Experiments

In biology many investigations use experiments of some sort. An experiment occurs if anything is altered or controlled by the investigator. For example, an investigation into the effect of fertilizer on plant growth will use a control plot (or several control plots) where there is no fertilizer added and then one or more plots where fertilizer has been added at known concentrations set by the investigators. In this way the effect of fertilizer can be determined by comparison of the different concentrations of fertilizer. The condition being controlled (e.g. fertilizer) is usually called a factor and the different levels used called treatments or factor levels (e.g. concentrations of fertilizer). The design of this experiment will be determined by the hypothesis or hypotheses being investigated. If the effect of the fertilizer on a particular plant is of interest then, perhaps, a range of different soil types might be used with and without fertilizer. If the effect on plants in general is of interest then an experiment using a variety of plants is required, either in isolation or together. If the optimum fertilizer treatment is required then a range of concentrations will be applied and a cost–benefit analysis carried out.

More details and strategies for experimental design are considered in Chapter 4.

Statistics

In general, statistics are the results of manipulation of observations to produce a single or small number of results. There are various categories of statistics depending on the type of summary required. Here I divide statistics into four categories.

Descriptive statistics

The simplest statistics are summaries of data sets. Simple summary statistics are easy to understand but should not be overlooked. These are not usually considered to be statistics but are in fact extremely useful for data investigation. The most widely used are measures of the 'location' of a set of numbers such as the mean or median. Then there are measures of the 'spread' of the data, such as the standard deviation. The choice of appropriate descriptive statistics and the best way of displaying the results are considered in Chapters 5 and 6.

Tests of difference

A familiar question in any field of investigation is going to be something like '*is this group different from that group?*'. A question of this kind can then be turned into a null hypothesis with a form: '*this group and that group are not different*'. To answer this question, and test the null hypothesis, a statistical test of difference is required. There are many tests that all seem to answer the same type of question but each is appropriate when certain types of data are being considered. After the simple comparison of two groups there are extensions to comparisons of more than two groups and then to tests involving more than one way of dividing the individuals into groups. For example, individuals could be assigned to two groups by sex and also into groups depending on whether they had been given a drug or not. This could be considered as four groups or as what is known as a factorial test where there are two factors 'sex' and 'drug' with all combinations of the levels of the two factors being measured in some way. Factorial designs can become very complicated but they are very powerful and can expose subtleties in the way the factors interact that can never be found through investigation of the data using one factor at a time.

Tests of difference can also be used to compare variables with known distributions. These can be statistical distributions or derived from theory. Chapter 7 considers tests of difference in detail.

Tests of relationships

Another familiar question that arises in scientific investigation is in the form '*is A associated with B?*'. For example 'is fat intake related to blood pressure?'. This type of question should then be turned into a null hypothesis that '*A is not associated with B*' and then tested using one of a variety of statistical tests. As with tests of difference there are many tests that seem to address the same type of problem, but again each is appropriate for different types of data.

Tests of relationships fall into two groups called correlation and regression depending on the type of hypothesis being investigated. Correlation is a test to measure the degree to which one set of data varies with another — it *does not* imply that there is any 'cause' and 'effect' relationship. Regression is used to fit a relationship between two variables such that one can be predicted from the other. This *does* imply a 'cause' and 'effect' relationship or at least an implication that one of the variables is a 'response' in some way. So in the investigation of fat intake and blood pressure a strong positive correlation between the two shows a relationship but does not show cause and effect. If there were a significant positive regression line, this would imply that blood pressure can be predicted using fat intake *or*, if the regression uses the fat intake as the 'response', that fat intake can be predicted from blood pressure.

There are many additional techniques that can be employed to consider the relationships between more than two sets of data. Tests of relationships are described in Chapter 8.

Tests for data investigation

A whole range of tests is available to help investigators explore large data sets. Unlike the tests considered above, data investigation need not have a hypothesis for testing. For example, in a study of the morphology of fish there may be many fin measurements from a range of species and sites that offer far too many potential hypotheses for investigation. In this case the application of a multivariate technique may show up relationships between individuals, help assign unknown specimens to categories or just suggest which hypotheses are worth further consideration.

A few of the many different techniques available are considered in Chapter 9.

Choosing a test: a key

I hope that you are not reading this section with your data already collected and the experiment or sampling programme 'finished'. If you have finished collecting your data I strongly advise you to approach your next experiment or survey in a different way. As you will see below, I hope that you will be using this key *before* you start collecting real data.

Remember — eight steps to successful data analysis

1 Decide what you are interested in.
2 Formulate a hypothesis or hypotheses.
3 Design the experiment, manipulation or sampling routine.
4 Collect dummy data. Make up approximate values based on what you expect.
5 *Use the key here to decide on the appropriate test or tests.*
6 Carry out the test(s) using the dummy data.
7 If there are problems go back to step 3 (or 2); otherwise collect the real data.
8 Carry out the test(s) using the real data.

The art of choosing a test

It may be a surprising revelation, but choosing a test is not an exact science. There is nearly always scope for considerable choice and many decisions will be made based on personal judgements, experience with similar problems, or just plain hunches. There are many circumstances where there are several ways that the data could be analysed and yet each of the possible tests could be justified.

A common tendency is to force the data from your experiment into a test you are familiar with even if it is not the best method. Look around for different tests that may be more appropriate to the hypothesis you are testing. In this way you will expand your statistical repertoire and add power to your future experiments.

A key to assist in your choice of statistical test

Starting at point 1 move through the key following the path that best describes your data. If you are unsure about any of the terms used then consult the glossary or the relevant sections of the next two chapters. This is not a true dichotomous key and at several points there are more than two routes or end points.

There may be several end points appropriate to your data that result from this key. For example you may wish to know the correct display method for the data and then the correct measure of dispersion to use. If this is the case, go through the key twice. All the tests and techniques mentioned in the key are described in later chapters.

Italics indicate instructions about what you should do.

Numbers in brackets indicate that the point in the key is something of a compromise destination.

There are several points where rather arbitrary numbers are used to determine which path you should take. For example, I use 30 different observations as the arbitrary level at which to split continuous and discontinuous data. If your data set falls close to this level you should not feel constrained to take one path if you feel more comfortable with the other.

---

1 Testing a clear hypothesis and associated null hypothesis   25
(e.g. H_1 = blood glucose level is related to age and H_0 = blood
glucose is not related to age).
Not testing any hypothesis but simply want to present, summarize   2
or explore data.

2 Methods to summarize and display the data required.   3
Data exploration for the purpose of understanding and getting a   60
feel for the data or perhaps to help with formulation of hypotheses.
For example, you may wish to find possible groups within the data
(e.g. 10 morphological variables have been taken from a large
number of carabid beetles; the multivariate test may establish
whether they can be divided into separate taxa).

3 There is only one collected variable under consideration (e.g. the   4
only variable measured is brain volume although it may have been
measured from several different populations).
There is more than one variable (e.g. you have measured the number   24
of algae per millilitre *and* the water pH in the same sample).

4 The data are discrete; there are fewer than 30 different values   5
(e.g. number of species in a sample).
The data are continuous; there are more than 29 different values   16
(e.g. bee wing length measured to the nearest 0.01 mm).
(Note: the distinction between the above is rather arbitrary.)

5 There is only one group or sample (e.g. all measurements taken   6
from the same river on the same day).
There is more than one group or sample (e.g. you have measured   15
the number of antenna segments in a species of beetle and have
divided the sample according to sex to give two groups).

6 A graphical representation of the data is required.   7
A numerical summary or description of the data required.   11

7 A display of the whole distribution is required.   8
A crude display of the position and spread of data is required —
use a box and whisker display to show medians, range and interquartile
range, p. 46 (also known as a box plot).

8 Values have real meaning (e.g. number of mammals caught per night).   10
Values are arbitrary labels that have no real sequence (e.g. different   9
soil type classifications in an area of forest).

9 There are fewer than 10 different values or classifications — *draw*
a pie chart, p. 49. Ensure that each segment is labelled clearly and that
adjacent shading patterns are as distinct as possible. Avoid using 3D or
shadow effects, dark shading or colour. Do not add the proportion in
figures to the 'piece' of the pie as this information is redundant.
There are 10 or more different values or classifications — *amalgamate*
values until there are fewer than 10 or divide the sample to produce two
sets each with fewer than 10 values. Ten is a level above which it is
difficult to distinguish different sections of the pie or to have sufficiently
distinct shading patterns.

10 There are more than 20 different values — *amalgamate values to*
produce around 12 classes (almost certainly done automatically by your
package) and draw a histogram, p. 47. Give the classes on the x-axis,
frequency of occurrence (number of times the value occurs) on the y-axis,
and no gaps between bars. Do not use 3D or shadow effects.
There are 20 or fewer different values — *draw a bar chart, p. 47.*
Each value should be represented on the x-axis. If there are few classes,
extend the range to include values not in the data set at either side; give
frequency of occurrence (number of times the value occurs) on the y-axis.
Gaps should appear between bars, unless the variable is clearly supposed
to be continuous; do not use 3D or shadow effects.

11 Measure of position (mean is the one used most commonly).   12
Measure of dispersion or spread (standard deviation and   13
confidence intervals are the most commonly used).
Measure of symmetry or shape of the distribution.   14
(Note: you will probably want to go for at least one measure of
position and another of spread in most cases.)

12 Variable is definitely discrete and usually restricted to integer values smaller than 30 (e.g. number of eggs in a clutch) — *calculate the median, p. 50.*
Variable should be continuous but has only a few different values due to accuracy of measurement (e.g. bone length measured to the nearest centimetre) — *calculate the mean, p. 49.*
If you are particularly interested in the most commonly occurring response — *calculate the mode, p. 50, in addition to either the mean or median.*

13 Very rough measure of spread is required — *calculate the range, p. 51 (note that this measure is very biased by sample size and is rarely a useful statistic).*
You are particularly interested in the highest and/or lowest values — *calculate the range, p. 51.*
Variable should be continuous but has only a few values due to accuracy of measurement — *calculate the standard deviation, p. 52.*
Variable is discrete or has an unusual distribution — *calculate the interquartile range, p. 51.*

14 Variable should be continuous but has only a few values due to accuracy of measurement — *calculate the skew* (g_1), *p. 53.*
Observations are discrete or if you have already calculated the interquartile range and the median — *the relative size of the interquartile range above and below the median provides a measure of the symmetry of the data.*

15 You have not established the appropriate technique for a single sample — *go back to 6 to find the appropriate techniques for each group. You should find the same is correct for each sample or group.*   (6)
The samples can be displayed separately — *go back to 7 and choose the appropriate style. So that direct comparisons can be made, be sure to use the same scales (both x-axis and y-axis) for each graph. Be warned that packages will often adjust scales for you. If this happens you must force the scales to be the same.*   (7)
The samples are to be displayed together on the same graph — *use a chart with a box plot for each sample and the x-axis representing the sample number, p. 57. Make sure there is a clear space between each box plot.*

16 There is only a data set from one group or sample.   17
The data have been collected from more than one group or sample (e.g. you have measured the mass of each individual of a single species of vole from one sample and have divided the sample according to sex).   23

17 A graphical representation of the data is required.   18
A numerical summary or descriptive statistics are required.   19

18 Display of the whole distribution is required — *group to produce around 12–20 classes and draw a histogram, p. 47 (probably done automatically by your package). Use classes on the x-axis, frequency of occurrence (number of times the value occurs within the class) on the y-axis, and no gaps between bars, no 3D or shadow effects. Even-sized classes are much easier for a reader to interpret. Data with an unusual distribution (e.g. there are some extremely high values well away from most of the observations) may require transformation before the histogram is attempted.*
A crude display of position and spread of the data is required — *the 'error bar' type of display is unusual for a single sample but common for several samples. There is a symbol representing the mean and a vertical line representing the range of either the 95% confidence interval or the standard deviation, p. 57.*

19 Measure of position (mean is the most common).   20
   Measure of dispersion (spread).   21
   Measure of symmetry or shape of the distribution.   22
   You wish to determine whether the data are normally distributed — *carry out a Kolmogorov–Smirnov test, p. 77, an Anderson–Darling test, p. 79 or a chi-square goodness of fit, p. 68.*
(Note: you probably require one of each of the above for a full summary of the data.)

20 Unless the variable is definitely discrete or is known to have an odd distribution (e.g. not symmetrical) — *calculate the mean, p. 49.*
If the data are known to be discrete or the data set is to be compared with other, discrete data with fewer possible values — *calculate the median, p. 50.*
If you are particularly interested in the most commonly occurring value — *calculate the mode, p. 50,* **in addition to** *the mean or median.*

21 If the data are continuous and approximately normally distributed and you require an estimate of the spread of data — *calculate the standard deviation (SD), p. 52.*
(Note: standard deviation is the square root of variance and is measured in the same units as the original data.)
If you have previously calculated the mean and require an estimate of the range of possible values for the mean — *calculate 95% confidence limits for the mean, p. 52 (a.k.a. 95% confidence interval or 95% CI).*
A very rough measure of spread is required — *calculate the range, p. 51. (Note that this measure is very biased by sample size and is rarely a useful statistic in large samples.)*
If you have a special interest in the highest and or lowest values in the sample — *calculate the range, p. 51.*

If the data are known to be discrete or are to be compared with other, discrete data or if you have previously calculated the median — *calculate the interquartile range, p. 51.*
(Note: many people use *standard error (SE)* as a measure of spread. I am convinced that the main reason for this is that it is smaller than either SD or 95% CI rather than for any statistical reason. Do not use standard error for this purpose unless you are making a comparison to previously calculated SEs.)

22 If the data are continuous and normally distributed and you require an unbiased estimate of the symmetry of the data — *calculate the skewness/asymmetry of the data* (g_1), *p. 53.*
If the data are continuous and normally distributed, you have calculated skewness and you require an estimate of the 'shape' of the distribution of the data — *calculate the kurtosis* (g_2), *p. 53. (This is rarely required as a graphical representation will give a better understanding of the shape of the data. Kurtosis is only really worth calculating in samples with more than 100 observations.)*
If you have already calculated the interquartile range and the median — *re-examine the interquartile range. The relative size of the interquartile range above and below the median provides a measure of the symmetry of the data.*

23 You have not established the appropriate technique for a single   (16)
sample — *go back to 16 to find the appropriate techniques for each of the groups. You should find the same is appropriate for each sample or group.*
The samples can be displayed separately — *go back to 17 and choose*   (17)
the appropriate style. So that direct comparisons can be made, be sure to use the same scales (both x-axis and y-axis) for each graph. Be warned that statistical packages will often adjust scales for you.
The samples are to be displayed together on the same graph — *use a chart with an 'error bar' (showing the mean and a measure of spread) for each sample and the x-axis representing the sample number/site. Do not join the means unless intermediate samples would be possible (i.e. do not join means from samples divided by sex or species but do join those representing temperature or nutrient concentration, if the intervals between different sample temperatures are even).*

24 If each variable is to be considered separately — *go back to 4 and*   (4)
consider each variable in turn.
Two variables only — *a 2D scatter plot can be drawn, p. 58. The choice of variable for the x-axis and y-axis is free but if you suspect a possibility of 'cause' and 'effect', the 'cause' should always be on the x-axis. Do not draw a line of best fit even if it is offered by the package unless the situation is appropriate and you have carried out a regression analysis.*
Three variables — *a 3D scatter plot can be drawn, p. 63. It is very*

difficult to represent three dimensions on a 2D sheet of paper or computer screen. You must drop spikes to the 'floor' or 'origin' of the graph, otherwise it is impossible to visualize the spread in the third dimension. It may be better to use a series of 2D scatters instead.
More than three variables — *use a series of 2D or 3D scatterplots, p. 63.*

25 (Note: the distinction here will be slightly fuzzy in some cases, but essentially there are two basic types of test.)
The hypothesis is investigating differences and the null hypothesis that there is no difference between groups or between data and a particular distribution (e.g. H_1 [hypothesis]: white-eye and carmine-eye flies have different development times; H_0 [null hypothesis]: white-eye and carmine-eye flies do not have different development times).   26
The hypothesis is investigating a relationship and the null hypothesis is that there is no relationship (e.g. H_1 [hypothesis]: plant size is related to available phosphorous in the soil; H_0 [null hypothesis]: plant size is not related to the amount of available phosphorus).   46

26 Data are collected as individual observations (e.g. height in centimetres).   29
Data are in the form of frequencies (e.g. when carrying out a plant cross and scoring the number of offspring of each type).   27

27 There are two possibilities only (e.g. white or pink).   28
There are more than two possibilities — *carry out a G-test, if your package supports it, p. 66, otherwise use a chi-square goodness of fit, p. 68.*
There are more than about eight possibilities — *a Kolmogorov–Smirnov test, p. 77, may be more convenient than the chi-square goodness of fit, p. 68.*

28 There are more than 200 observations in the sample — *carry out a G-test, p. 66, if your package supports it, otherwise use a chi-square goodness of fit, p. 68.*
There are 25–200 observations — *carry out a G-test if your package supports it, p. 66, otherwise use a chi-square goodness of fit, p. 68, but add a 'continuity correction' by adding 0.5 to the lower frequencies and subtracting 0.5 from the higher. This is very conservative and may result in a non-significant result when a marginally significant one is present (type I error). If your package supports the 'Williams correction' then use that instead of the 'continuity correction'.*
Fewer than 25 observations — *four possible solutions in order of preference: use a binomial test if supported by your package; carry one out by hand if you are able; get a bigger sample; pretend you have 25 observations and use the instructions above.*

29 There is only one way of classifying the data (e.g. grouped by species).   30
    There is more than one way of classifying the data (e.g. grouped by   38
    species *and* collection site).

30 There are only two groups (e.g. male and female or 'before' and   31
    'after').
    There are more than two groups (e.g. samples from four different   35
    fields).
    (Note: the null hypothesis is that all groups have the same mean so
    if any two groups have different means you have to reject this null
    hypothesis.)
    There are more than two groups and several measured variables
    (e.g. individuals divided by species [a grouping variable] and the
    measured variables are various anatomical characters or dimensions
    such as leaf length, stem thickness and petal length) — *canonical*
    variate analysis, p. 205.

31 Two samples are 'paired'. This means that the same individual,   32
    location or quadrat has been measured twice. This is the 'before'
    and 'after' design (e.g. river nitrate level is measured before and
    after a storm).
    Two samples are independent. There are different groups of   34
    individuals in the two samples.

32 The data are normally distributed, there are at least 30 possible values
    and variances are, at least approximately, homogeneous — *carry out a*
    paired t-test, p. 80. To test for normal distribution use the Kolmogorov–
    Smirnov test, p. 77, Anderson–Darling test, p. 79, or chi-square goodness
    of fit, p. 68. A test for homogeneity of variance is often an option within
    the t-test *in the package (e.g. a Levene test or Bartlett test).*
    A two-way ANOVA *test is a potential alternative here but is more*
    difficult to carry out than the paired t-test *in most statistical packages,*
    p. 107 (use one factor of the ANOVA *to represent 'before/after' and the*
    other to represent the different individuals).
    Above does not, or might not, apply.   33

33 Data have more than 20 possible values — *carry out a Wilcoxon*
    signed ranks test, p. 83.
    Data have 20 or fewer possible values (e.g. questionnaire results
    with a question of 'how do you feel' asked before and after exercise)
    — *carry out a sign test if supported by your package (this is a very*
    conservative but fairly low power test), p. 85. If this is not available
    in the package carry out a Wilcoxon signed ranks test, p. 83.

34 The data set is normally distributed, there are at least 30 possible
    values and variances are, at least approximately, homogeneous — *carry*
    out a one-way ANOVA *with one factor having two levels (one for each group),*

p. 94, or use a t-test, *p. 89. To test for normal distribution use the Kolmogorov–Smirnov test, p. 77, Anderson–Darling test, p. 79, or chi-square goodness of fit, p. 68. A test for homogeneity of variance is often an option within the* t-test *or the* ANOVA *in the package (e.g. a Levene test). The traditional method is to use a* t-test *for this type of experiment but it is no better than* ANOVA *in this circumstance as both tests give an identical result. Many packages have versions of the* t-test *that make an adjustment to the degrees of freedom to account for violations of the assumptions of the test.* The data set does not, or might not, fulfil the conditions above — *carry out a Mann–Whitney U test, p. 101 (sometimes called Wilcoxon–Mann–Whitney or Wilcoxon two-sample test* **not** *Wilcoxon signed ranks test). (The Kruskal–Wallis test is an alternative but is less powerful.)*

35 Samples are 'repeated measures'. This means that the same individual is measured through time. This is an extended 'before' and 'after' design (e.g. lake turbidity is measured each year for several years). **36**
Each sample is independent. There are different groups of individuals in each sample. It is important that no individual is present more than once in the data set, otherwise problems (of inappropriate replication) reduce the power of the statistical test. **37**

36 The data set is normally distributed, there are at least 30 possible values and variances are, at least approximately, homogeneous — *carry out a two-way, repeated measures* ANOVA *with one factor having a different level for each sampling repeat and a second factor having a level for each individual you are sampling, p. 107 (easy if you have only five rivers measured each year but very tedious to input and difficult to interpret if you have 50). Be aware that if your package does not support repeated measures designs the degrees of freedom in a two-way* ANOVA *should be reduced to compensate for the design. To test for normal distributions you can use the Kolmogorov–Smirnov test, p. 77, Anderson–Darling test, p. 79, or chi-square goodness of fit, p. 68, although in practice it is usual to use experience to determine whether the data are likely to be normally distributed. Furthermore the* ANOVA *is quite robust to small departures from a normal distribution.*
The data set does not conform to the restrictions above and you only have one observation for each repeat of each sample — *carry out a Friedman test with one factor having a different level for each sampling repeat event, p. 122 (e.g. before, during, after), and one factor having a different level for each individual (e.g. person) you are sampling.*
Neither of the above apply. This one is difficult! It results from poor planning of the experiment — *usually it is best to carry out an* ANOVA, *p. 137, as if the data conformed to the assumptions of distribution and variances but to treat the resulting P-values with caution especially if the P-value is between 0.1 and 0.01.*

37 The data set is normally distributed, there are at least 30 possible values and variances are, at least approximately, homogeneous — *carry out a one-way ANOVA with the one factor having one level for each group, p. 109. (Note: the t-test can only be used on two groups.) If the result is significant then you need to carry out a post hoc test to determine which factor levels are significantly different from which (p. 116). If you are cautious, or unsure, use a Kruskal–Wallis test instead, p. 119.*
The data set does not, or might not, fulfil the conditions above — *carry out a Kruskal–Wallis test with one factor having a level for each group, p. 119. (Note: the Mann–Whitney U test only works for two groups so is not appropriate here.)*

38 There are only two factors/ways of classifying the data (e.g. strain and location).   39
There are three factors/ways of classifying the data (e.g. sex, region and year).   43
There are more than three factors — *use the key as if there are three factors and extrapolate. Multifactorial experimental designs become increasingly difficult to interpret because there are so many possible interactions between factors and it is often easiest to leave out factors that you have proved to have no significant effect, p. 157.*   (43)

39 There is no replication (i.e. only one value assigned to each combination of the two factor levels) (e.g. basal trunk diameters are collected from four strains of apple tree grown under four watering regimes after 2 years but with only one tree under each watering condition).   40
There is replication (i.e. there are two or more values for each combination of the two factors).   41

40 The data are likely to be normally distributed (it is impossible to test this when there is only one observation in each factor combination). Data such as lengths and concentrations are likely to be appropriate but judgement is required — *carry out a two-way ANOVA, p. 126, but note that you will not be able to look for any interaction between the two factors.*
You are cautious or have a data set that is unlikely to be normally distributed — *carry out a Friedman test, p. 122.*

41 Factors are fully independent of each other.   42
One factor is 'nested' within another (e.g. if there are three branches sampled from each of three trees then the branch is said to be 'nested' within the trees) — *carry out a nested ANOVA, p. 160 (a.k.a. hierarchical ANOVA). (Note: there is no non-parametric [i.e. one that makes fewer assumptions about the distribution of the data] equivalent of this test.)*

42 The data set is normally distributed, there are at least 30 possible values and variances are approximately equal — *two-way ANOVA,*

p. 137, measure of the interaction between the two factors is possible.
The data set is not as above. Versions of a two-way ANOVA making
fewer assumptions about the data (i.e. non-parametric) are a fairly
recent innovation and are not yet appearing in statistical packages.
If the experiment is balanced or nearly so (i.e. there are the same
number of observations for each combination of factor levels) —
*carry out a Scheirer–Ray–Hare test, p. 145. This will, almost certainly,
not be in your statistical package but can still be carried out with a little
modification of other tests. See the section describing the test for details.*

43 (Note: there are no non-parametric tests available from here on so
if the data set does not fit the assumptions of the test you have no
alternatives. ANOVA is quite robust to failure to meet its assumptions
but be aware, especially if results are borderline.)
All factors (ways of grouping the data) are independent of each other.　　44
At least one factor is nested in another (e.g. in an experiment the　　　45
variable is blood sugar level in mice. The factors are litter, female
and food provided. If there are two litters from each of two females
then litter will be 'nested' within female. Neither litter nor female
will be 'nested' within food.)

44 There is only one observation for each combination of factor levels
— *carry out a three-way ANOVA, p. 151. You will not be able to calculate
the significance of the three-way interaction but you will be able to do
this for the interaction between each combination of two factors. (Note
that any main factors that prove to be non-significant can be left out
of the analysis to reduce the complexity of the design.)*
There is more than one observation for each combination of factor
levels — *carry out a three-way ANOVA, p. 151; interaction terms are possible.*

45 One factor is 'nested' within another and the third is independent
(as in the mouse example in 43) — *carry out an ANOVA involving
both hierarchical and crossed factors. This is often difficult to reach in
statistics packages although the design is a common one. If you only have
one observation for each combination of factor levels then an interaction
term cannot be tested (this is because it has to be used as the residual or
error term).*
One factor is 'nested' within another that is itself 'nested' within
a third (e.g. in a water pollution survey the variable is nitrate
concentration. Several samples have been taken from five streams
from each of three river systems and this has been done in two
countries. The factors are 'stream', 'river' and 'country'. 'Stream'
is nested within 'river' and 'river' nested within 'country') —
carry out a nested or hierarchical ANOVA, p. 160.

46 (The choice you have here is one that is frequently confused,
be careful.)

The purpose of the test is to look for an association between variables (e.g. is there an association between wing length and thorax length?). You have not set (controlled) one of the variables in the experiment. There is no reason to assume a 'cause' and 'effect' relationship. This is a test of correlation.   47

One or more of the variables has been set (controlled or selected) by the experiment or there is a probable 'cause' and 'effect' or functional relationship between the variables. One of the uses of the regression statistics you are moving to is prediction (e.g. the experiment is looking at the effect of temperature on heart rate in *Daphnia*. You are expecting that heart rate is affected by temperature but wish to discover the form of the relationship so that predictions can be made). This is a regression type of test.   53

47 Data are in the form of frequencies (e.g. number of 'white flowers' and 'orange flowers').   48

There is a value for each observation. Variables should be paired, etc. (e.g. an observation of two variables 'cell count' and 'lung capacity' from one individual).   50

48 There are two variables — *if you follow this thread further you will reach tests that are often awkward to carry out in packages and are often easier to calculate by hand. If you do calculate them by hand you may have to look up the significance level using a chi-square table.*   49
There are more than two variables — *simultaneous comparisons of frequencies for more than two classifications are very difficult to interpret. It is best to compare them pairwise.*

49 The two variables each have two possible values (e.g. 'yes'/'no' or 'male'/'female') — *calculate a phi coefficient for a 2 × 2 table, p. 172, if your package supports it or you can do it by hand. This test is a special case of a contingency chi-square calculation, p. 165.*
At least one of the variables has more than two possible values (e.g. a crude land classification 'forest'/'scrub'/'pasture'/'arable' is compared to an estimate of the density of a small mammal 'common'/'rare'/'absent') — *calculate a contingency chi-square, p. 165, and, if your statistical package supports it, a Cramér coefficient, p. 172.*

50 There are two variables.   51
There are more than two variables.   52

51 Both sets of data are continuous (have more than 30 values) and are approximately normally distributed (a good way to get a feel for this is to produce a scatterplot which should produce a circle or ellipse of points) — *carry out a Pearson product–moment correlation, p. 173 (coefficient is called* r*). This is the standard correlation method.*

Data are discrete, or not normally distributed, or you are unsure — *use a Spearman rank-order correlation coefficient, p. 177 or a Kendall rank-order correlation coefficient, p. 179. The marginal advantage of the former is that it is slightly easier to compare with the Pearson product–moment correlation, while the latter can be used in partial correlations.*

Data are ranked — *use a Kendall rank-order correlation coefficient, p. 179. (The Spearman correlation is marginally inferior in this case.)*

52 (Note: partial and multiple correlations are difficult to interpret.) All sets of data are continuous and approximately normally distributed and you are interested in the direct level of association between pairs of variables — *use pairwise measures of association using a Pearson correlation, p. 173.*

All sets of data are continuous and approximately normally distributed, and you are interested in the overall pattern of association — *use partial correlation, p. 194, which looks at the correlation between two variables while the others are held constant. (Multiple correlation is a possibility but is rarely supported in packages. Its disadvantage is in interpretation and its inability to distinguish positive and negative relationships.)*

Above does not apply, or you are cautious — *carry out the Kendall partial rank-order correlation coefficient, p. 194, a test that finds the correlation between two variables while a third is held constant. This may not be supported by your package. If it is not, pairwise testing is the only alternative.*

53 The dependent variable is discrete, not normally distributed or ranked. Be warned that non-parametric regression is required and that this is rarely available in a statistical package.   54

The dependent 'effect' variable is continuous and at least   55
approximately normally distributed with the same variation in 'effect' for any given value of the 'cause' variable. (There will often be a requirement for a transformation of the data. Proportions and percentages can be transformed using the arcsine transformation (p. 41) or probits. Other distributions may be normalized using reciprocal transformations or many other possibilities. It is important that efforts are made to fulfil the requirements for approximately normal data with equal variance using transformations.)

The dependent 'effect' variable is a proportion or frequency (e.g. proportion of population with a disease). The 'cause' variable is measured without error and is chosen or set by the experimenter — *use logistic regression, p. 189.*

54 There is one independent 'cause' variable and one dependent (55)
'effect' variable — *use the Kendall robust line-fit method. If this is
not available consider reframing (usually by simplifying) your
hypothesis somewhat to fit a non-parametric correlation. The only
other alternative is to continue to a parametric test (55) being very
cautious with interpretation of the results.*

All other designs — *there is no satisfactory non-parametric test* (55)
*and certainly nothing in a statistical package yet. Either reframe the
hypothesis or go to 55 and continue with a parametric test. If there
are two 'cause' variables and one 'effect' then the 'cause' variables
might be divided into a small number of categories (e.g. 'low', 'medium'
and 'high') and then a Scheirer–Ray–Hare test carried out, p. 145.*

55 There is one dependent variable ('effect') and one independent 56
variable ('cause').

There is one or more dependent variable ('effect') and two 58
or more independent variables.

The data for the dependent variable can be classified into more
than one group (e.g. by 'species' or 'sex'). There is a variable that
may affect the dependent variable — ANOVA *(analysis of covariance)
is required, p. 195. This is a technique where the confounding variable,
known as the covariate, is factored out by the analysis allowing
comparison of the groups. Complex designs are possible but the most
common is analogous to a one-way* ANOVA *with the data (e.g. dry
weight) in classes (e.g. cultivars) and a variable known to be
confounding factored out as the covariate (e.g. degree days).*

56 The independent 'cause' variable is measured without error. 57
There is known to be some measurement error associated with the
independent variable — *a model II regression is required, p. 192, or
Kendall robust line-fit method, p. 189. This is a rarely used technique
and only occasionally appears in statistical packages. It has the odd
property of always overestimating the slope of the relationship compared
to the result from a normal (model I) regression, p. 182.*

57 (Note: as the theoretical shape of the relationship is often unknown
the usual strategy here is to try both methods and see which gives
the better fit.)

The relationship is likely to be a straight line or you are not sure of
the form of the relationship — *linear regression, p. 182 (a.k.a. model
I regression). (Note: in many cases the independent variable can be
transformed to straighten the relationship between cause and effect
(e.g. if the independent variable is 'size' and is right skewed then a
log transformation will often improve a linear fit.)*

The relationship is curvilinear or complex — *polynomial regression or
quadratic regression (a special case of polynomial regression), p. 192.*

58 There is one dependent 'effect' variable and two or more
   independent 'cause' variables.
   There are several 'cause' and 'effect' variables — *use path analysis,
   p. 199.*

59

59 Your primary aim is to find the 'cause' variable(s) that are the best
   predictors of the 'effect' variable — *use stepwise regression, p. 198.*
   You want to establish a model using all available 'cause' variables
   — *use multiple regression, p. 198.*
   (*Note*: the distinction between these two is rather arbitrary.)

60 You have arrived at principal component analysis, discriminant
   function analysis and other multivariate techniques for exploring your
   data. The usual result of this type of exploration is to identify simple
   relationships hidden in the mass of the data. Some of these tests are
   described in Chapter 9.
   There are several observed variables that are approximately
   continuous and you have no preconceived notion about division
   into groups — *use principal component analysis, p. 200.*
   There are a variety of variables that may be a combination of 'causes'
   and 'effects' — *use path analysis, p. 199.*
   There are two or more sets of observations and one or more grouping
   variables — *use* MANOVA *(multivariate analysis of variance), p. 210.*
   There are two or more sets of observations, one or more grouping
   variables and a recorded variable that is known to affect the observed
   variables (e.g. temperature) — *use* MANCOVA *(multivariate analysis of
   covariance), p. 212.*
   There are several observed variables for each individual that are
   approximately continuous and individuals have already been assigned
   to groups (e.g. species) — *use canonical variate analysis, p. 205.*
   There are several observed variables for each individual that are
   approximately continuous, individuals have already been assigned to
   groups (e.g. species) and the intention is to assign further individuals
   to appropriate groups — *use discriminant function analysis, p. 205.*
   There are several observed variables for each individual that are
   categorical or nominal, individuals have already been assigned to
   groups (e.g. species) and the intention is to assign further individuals
   to appropriate groups — *use logistic regression, p. 189.*
   There are several observed variables for each individual and you wish
   to determine which individuals are most similar to which — *use cluster
   analysis, p. 212.*
   You have data on the relative abundance of species from various sites
   and wish to determine similarities between sites — *use cluster analysis,
   p. 212, or* TWINSPAN, *p. 215.*

4

Hypothesis testing, sampling and experimental design

This chapter expands on some of the ideas introduced in Chapter 2.

Hypothesis testing

Much of scientific investigation is based on the idea of hypothesis testing. The idea is that you formalize a **hypothesis** (H_1) into a statement such as '*male and female shrimps are different sizes*', collect appropriate data, and then use statistics to determine whether the hypothesis is true or not.

However, it is not quite as simple as that. The statistical tests do not give a simple answer of true or not. First you have to realize that every hypothesis will have an associated **null hypothesis** (H_0) and most statistical tests use the null hypothesis as a starting point.

So, for this example, the hypothesis (H_1) is: '*male and female shrimps are different sizes*' and the associated null hypothesis (H_0) is: '*male and female shrimps are **not** different sizes*'.

What a statistical test determines is the probability that the null hypothesis is true (called the *P*-value). If the probability is low then the null hypothesis is rejected and the original hypothesis accepted.

Acceptable errors

In reality, the null hypothesis is either true or false. Unfortunately, we only have a sample from all the individuals in a population and the statistical test only gives a *probability* of the null hypothesis being true based on the sample. There are two ways of making the wrong inference from the test. These two types of errors are usually called **type I** and **type II errors**.

		Null hypothesis	
		Accepted	Rejected
Null hypothesis	True	Correct	**Type I error**
	False	**Type II error**	Correct

In a **type I error** the null hypothesis is really true (male and female shrimps are not different sizes) but the statistical test has led you to believe that it is false (there is a difference in size). This type of error is very dangerous and could be seen as a 'false positive'.

In a **type II error** the null hypothesis is really false (male and females are really different sizes) but the test has not picked up this difference. Small sample sizes will often lead to a type II error. This type of error is probably less dangerous than the type I but should still be avoided if possible.

The ideal statistical test should have an equal, and hopefully very low, chance of the two types of errors. A test which increases the chance of getting a type II error while decreasing the chance of a type I is said to be 'conservative' while one that increases the chance of a type I error is said to be 'liberal'. Although it is best to achieve this balance of type I and type II errors, a cautious approach is to err towards 'conservative' tests.

P-values

Errors are the inevitable consequence of results based on probability. The lower the probability (*P*-value) the more confident you can be in the rejection of the null hypothesis, but you can never be totally sure, unless you have measured the whole population, that you are correct. It is a usual convention in biology to use a critical *P*-value of 0.05 (often called alpha). This means that the probability of a null hypothesis being true is 0.05 (5% or one in 20). Whenever a statistical test gives a result with a *P*-value less than 0.05 we reject the null hypothesis and accept the alternative hypothesis.

There is nothing magical about $P < 0.05$; it is just a convention. If you use a lower critical *P*-value then the chance of making a type II error is increased. If you choose a higher critical *P*-value then you increase the chance of making a type I error.

It is worth pointing out that if a *P*-value is less than 0.05 it does not *prove* that the null hypothesis is false, it just indicates that it is unlikely to be true. Indeed statistics can never *prove* anything, it can only suggest that a hypothesis is very likely to be true or untrue. It is also worth noting that a *P*-value above 0.05 certainly does not prove that the null hypothesis is true — it just indicates that there is not enough evidence to reject it.

It is conventional to indicate degrees of significance using asterisks in tables, or sometimes on figures. A single asterisk is usually used for *P*-values between 0.05 and 0.01, two asterisks for values below 0.01, and three asterisks either for results below 0.005 or 0.001. If asterisks are used in this way they should be decoded in a figure or table caption.

One final point about *P*-values is that when more than one test is used the critical *P*-value used should be reduced to retain a critical level of 0.05 across an experiment. This makes good sense as any experiment including 20 statistical tests should, on average, generate one significant result even when there is no biological effect. The two main methods for adjusting *P*-values to retain the experiment-wide threshold are discussed in more detail in the section on correlation (p. 193).

Sampling

Nearly all statistical tests make a fundamental assumption that sampling of individuals will be at random from all the individuals that could possibly be sampled. This sounds simple but achieving a random sample is not always easy.

In almost all biological studies it will be impossible to account for every individual in a population. Therefore it is necessary to examine a subgroup of the total population and extrapolate from this to the whole population. The process by which the subgroup of the population is selected is sampling.

If a population is evenly distributed through a habitat then a single small sample would be enough to gain a good estimate of whatever it is you are interested in (e.g. total population size, mean age or weight). However this is rarely the case and most populations have distributions that are either random or clumped. In such populations a single sample is unlikely to produce a good estimate of population size or mean height or the variance of leaf thickness.

There are a wide variety of sampling strategies in use. It is important to choose a strategy that is appropriate to the population being investigated. There are several steps in the development of a sampling strategy.

Choice of sample unit

A sampling unit may either be defined arbitrarily, such as a quadrat, transect or pitfall trap, or be defined naturally, such as a leaf or individual. Usually, naturally defined sample units will be obvious but the choice of an arbitrary unit size may be important. If the unit is a quadrat then there will obviously be a trade-off in effort between the number of sample units that can be observed and their size, simply because it takes more time to get information from a large quadrat.

A sample unit might also be a length of time (e.g. if you are investigating pollination then the number of times a flower is visited in a series of time periods of set length might be your data set).

> The size of sample unit will usually remain constant but may sometimes be variable, especially if the characteristics of a population distribution are being investigated. Methods using variable sample unit size to investigate distribution pattern are called quadrat-variance methods. These methods allow an observer to gain an insight into patterns of distribution in space or time by analysing the different characteristics of the samples (e.g. mean and variance) using different sample units.

Number of sample units

This is nearly always determined by the amount of labour available: the more time and people that are available the more information can be collected. However, it is possible to calculate the number of sample units required to produce an accurate estimate of the population size. In general more sample units will be preferred as the number of sample units increases the accuracy of statistical tests. Quantity should not be increased at the expense of quality. Poor quality data will have more inherent error and therefore make the statistics less powerful.

If you require general advice on the number of observations to make then I can only suggest that, as a rule of thumb, you need at least 20 observations for a sample using a measured variable and many more than that if the variable is a simple categorical one.

Positioning of sample units to achieve a random sample

An unbiased estimate of a population is only possible if the sample units are representative of the total population. The easiest way of achieving this is for each sample unit to contain a random sample of the population under investigation. If quadrats are being used then their position within the area under investigation should be chosen using random numbers to generate two coordinates that are then used to position a corner of the quadrat. Although this method of choosing a position using random numbers often requires an area to be marked out, it is to be preferred over quasirandom techniques, such as throwing a quadrat, that are certain to introduce some involuntary observer bias.

Selecting random individuals in an area for study can be difficult. Imagine a typical scheme for locating random plants: random coordinates are chosen and the nearest plant to the random location is selected for study. This apparently makes a random choice of individual plants, but in reality it introduces bias as isolated plants are much more likely to be selected than plants in the middle of clumps. Indeed the only way to really select individuals at random is, rather impractically, to label and number every individual and then select randomly from that list.

Random walks are another way to sample at random without requiring the area to be marked out. The observer walks a number of paces determined by a random number and then makes an observation or places a quadrat. Then another random walk is taken before the next observation, and so on. The advantages of this method are that sampling can be very rapid and that it requires little preparation. The drawback is that this type of sampling may be severely biased by the observer.

True random samples are ideal in a perfectly homogeneous habitat, but in a heterogeneous habitat they are likely to produce a biased sample with an estimated variance greater than that of the total population. A simple method used to minimize this problem is to take a **stratified random sample**. The method is simple: the total area is divided into equal plots and an even number of sample units is taken at random from each plot. It is possible to divide the total area into plots of different sizes if there are known to be different habitats in the total area. In this second case the number of sample units from each plot should be proportional to its size.

It might be tempting to conduct systematic sampling with sample units placed at regular intervals across a study area. There is a statistical problem with this strategy as most statistics require that a sample is taken at random from a population. However, some field ecologists suggest that estimates derived from systematic sampling are, on average, better than those from random sampling.

Timing of sampling

Most populations will be affected by season, time of day and local weather conditions. It is very important that timing is taken into account either by sampling strategy or by later analysis.

What I have been considering here is the problem faced by an ecologist working in the field and trying to design a suitable sampling strategy. The use of the very powerful statistical technique — analysis of variance (ANOVA) — is more common in the situation of a controlled experiment where you are analysing the effects of different levels of a treatment (e.g. concentrations of fertilizer or temperature) on some measured aspect of a population. Then, to get a true estimate of the effect of the treatment, experimental design will be of paramount importance.

Experimental design

I do not intend to say very much about experimental design here as there are whole books dedicated to the subject. However, that should not imply that experimental design is an uninteresting or unimportant subject. The appropriate design of an experiment is the key to successful analysis of a problem, for without the correct design you will never have the right sort of data. The problems of sampling still exist in a designed experiment but the control of the system allows the experimenter to ensure that there are sufficient individuals to sample, and that all factor combinations have the same number of observations.

Control

The idea of control in experiments is to remove the effect of all other factors apart from the one that is being investigated. The word is often applied to a group that has not been altered by an experimental manipulation and is used to compare with a group that has. The assumption is that everything apart from the manipulation is the same in the two groups, so any differences must be the result of the manipulation. There are different ways of applying controls and different types of control. It is important to consider whether the control used in an experiment is adequate to convince a sceptical reader that the effect 'proved' by the statistics is real or not. As a general rule of thumb more control is required! It is always tempting to focus on the more interesting manipulated groups and not give enough attention to control.

Procedural controls

These are often overlooked in experimental designs. The idea is that everything done to the manipulated group, apart from the actual treatment, is done to a procedural control group that is the same size as the experimental group and the untouched control group. The idea of procedural control groups has been widely used in recent medical studies and has shown some interesting results. For example, a new supplementary treatment for a common disease is being investigated. Everyone in the study is given the conventional treatment. Each individual is then randomly (and secretly) assigned to one of three categories: the control group is given nothing else, the experimental group is told about the supplementary treatment and given the new drug and the procedural control group is treated exactly as the experimental group except they are given a dummy drug (e.g. chalk tablet or water injection). In this way the effect of the drug can be differentiated from the effect of the procedure.

Procedural controls are especially important if the experimental system requires a lot of preparation through building exclosure fences or with repeated visits to a site or many interventions in a laboratory population. The technique should nearly always be used in conjunction with the 'untouched' control.

Temporal control

This is another aspect of experimental design that is worth incorporating. If the effect of a long-term manipulation is to be considered and there is only one control group and one experimental group available it is better to start the manipulation *after* the monitoring process has been underway for some time. The reason for this is that the differences between the two populations *without* any manipulation should be accounted for before the differences following the manipulation are tested. The ideal experiment will use half of the time for before and half for after manipulation. For example, if there are two lakes available to study the effect of eutrophication (surplus of nutrients) then the best design for a 2-year study is to monitor the lakes untouched for the first year and then to add nutrients to one of the two lakes during the second year.

Experimental control

This is any control of environment imposed by the experimenter. This is the classic type of control and is properly employed to remove all possible effects on the observations other than that from the experimental manipulation itself. Control as many factors as possible is always the best advice. So if the effect of

CO_2 levels on plant growth is being investigated then the experiment should control all the factors that may affect growth: light, temperature, humidity, water availability, soil organisms, soil type and nutrient availability. The degree of environmental control required to isolate the effect of the one factor being manipulated often leads to a very artificial situation with organisms being kept in isolation in perfect conditions. These controlled environments are often so far removed from the real world that the results are not really very informative.

Experimental control can be very expensive, requiring growth cabinets, controlled temperature chambers or incubators for even rather simple investigations.

Statistical control

This is an alternative to experimental control. Instead of fixing all the possible factors that can affect the observations they are measured instead. Careful recording of all the environmental conditions, both biotic and abiotic, that are known to affect the observations being collected can then be used in statistical analysis of the data. Providing the experiment is not confounded (e.g. if all the manipulated individuals are in a cold area and all the unmanipulated in the warm) it is often possible to unpick the various effects and remove them from the analysis to leave only the effect of the manipulation. If statistical control is to be attempted then efforts should be made to ensure that adequate monitoring of all the possible effects is carried out and that the individuals in the experimental and control groups experience a range of conditions.

Statistical control is usually cheaper than experimental control but requires more effort on the part of the researcher.

Some standard experimental designs

The **Latin square** is a system for placing replicates of treatments so that each of the treatment levels experiences each column and row of the experimental area. The reason for doing this is to avoid confounding the effect of the experimental treatment with any other factor that might be present in the experimental area (e.g. a gradient of soil quality). The arrangements for the four treatments suggested here is just one of many possible arrangements. Any arrangement of treatments such that each appears once in each column and row is alright, although it is probably best, as here, to have each treatment level only occurring in one corner.

A	B	C	D
C	A	D	B
D	C	B	A
B	D	A	C

If the experiment is carried out in a series of locations (often called blocks in statistical jargon) it is important to ensure that each of the treatment levels is equally represented in each of the blocks, otherwise any difference in conditions will be confounded with the treatment levels being investigated. Furthermore the position of the treatment levels within the blocks must not be repeated.

Block 1

A	C
D	B

Block 2

B	A
C	D

Block 3

D	B
C	A

Block 4

C	D
A	B

If a large number of samples are to be assigned to different treatment levels there are three obvious ways of assigning the levels. First to do all of level one, then two and so on; second to carry out the assignments entirely at random; and third to keep cycling through the levels in sequence. Each of these has problems. The first method will confound any external changes with the different treatment levels (e.g. if the experimenter becomes more efficient during the process). The second method is appealing but often leads to unwanted runs of the same treatment or too few replicates of particular treatments. The third may also confound the treatment levels with an external influence. The best strategy is a combination of the second and third methods and is called **stratified random** assignment. The assignments are made in batches with each treatment level appearing an equal number of times in the batch (usually one or two) but assigned at random. In the example shown there are three treatment levels (X, Y and Z) assigned twice each in five batches of six.

Batch 1	Batch 2	Batch 3	Batch 4	Batch 5
X	Z	Y	X	Z
Z	X	Y	Z	X
X	Y	Z	Y	X
Y	X	X	Z	Z
Y	Z	Z	Y	Y
Z	Y	X	X	Y

5

Statistics, variables and distributions

There are many books available that discuss the history, philosophy and workings of statistics at length. That is not my purpose here, but it is important to have at least some idea of what statistics are, how different statistics are appropriate in different circumstances and that there are different types of data that you might collect. This chapter covers much of the same ground as Chapter 2 but in much more detail. However, I am still only scratching the surface here and this section should only be used as a set of notes or pointers to further investigation of these subjects.

What are statistics?

In biology we are often concerned with groups of individuals. These 'individuals' might be single insects but they could also be, for example, herds, or species, or blood cells. In most cases it is totally impractical to measure every individual in the group or groups we are interested in. What we are forced to do instead is to take measurements from a subset of the group. We call these subsets of the whole group **samples**.

We can ask and answer questions about the groups by formulating hypotheses. A simple question could be: 'is species A bigger than species B?'. If we had access to data from all individuals in a group we could answer this type of question very easily. However, we only have the sample and from the sample we have to extrapolate to the whole group. This is the job of statistics.

For example, if the hypothesis is that the mean sizes of populations of pike in two lakes are different we could easily find the answer if we measured all the fish. However, in reality we only have a sample of 20 fish from each lake and the means of those samples might not be the same as that of all the fish. We carry out statistics on the information we have in the two samples to determine the probability that the hypothesis, or more usually its associated null hypothesis, is true. The idea of hypothesis formulation and testing has already been discussed in Chapter 4.

Types of statistics

I intend to say as little as possible about types of statistics here. However I feel it is important to give a feel for the differences and the way that statistics have been traditionally divided.

Descriptive statistics

These are usually the first to be calculated. They give information about the data you have collected. This can be a measure of the 'position' of the data, i.e. mean or median, and the 'dispersion', i.e. how variable the data are. Some descriptive statistics such as means (averages) will be familiar to everyone.

There is a division of statistics into two groups that are usually labelled 'parametric' and 'non-parametric'. This distinction is very real for statisticians but for those of us just using the tests it seems rather artificial.

Parametric statistics

These statistics make assumptions about the form of the data under investigation. For instance, they usually require variables to follow known distributions, usually the normal. If the data do conform to the assumptions then these tests are usually more powerful and should therefore be preferred. There are also types of questions that can only be answered if assumptions about distributions are made.

Non-parametric statistics

These are statistics that require little or no knowledge of the distribution of the data. Therefore they are often called 'distribution free', 'ranked' or 'ranking' tests. In general these tests are less powerful but 'safer' if you have not tested all the assumptions for a parametric test. Non-parametric tests are also somewhat restrictive and cannot be used to answer some more complicated questions.

In this book, unlike many other books, the chapters are not ordered according to the type of statistics. I have used the type of question you want to ask as the method of dividing up the book.

What is a variable?

To carry out any statistics you need some data to work with. First you decide what it is you are interested in and then select a suitable variable. The variable is the property that you measure. It is the food of statistics and choosing variables is something you must get right. For example, if you are interested in the occurrence of a scale insect on two strains of citrus trees then a suitable variable might be: 'number of scale insects on a leaf'. However, if the strains differ in leaf density or size of leaf then a more appropriate variable could be 'number of scale insects per square centimetre of leaf'. The insects may be bigger on one strain than the other so perhaps 'mass of scale insects per square centimetre of leaf' would be better. Each time the variable is refined in this example it becomes more difficult to obtain, taking more time and effort. There is a trade-off though: the more effort required for each observation leads to fewer data in total so any refinements to the variable collected must be warranted.

Choosing the best variable is something of an art.

It is important to ensure that the variable or variables you choose to measure or collect are appropriate to the task.

> *Note*: I use the term variable throughout this book as it is the one in common usage although the correct term is variate.

Types of variables or scales of measurement

There are many types of variable.

Measurement variables

These are variables where a numerical value is assigned. They can be further subdivided.

Continuous variables (sometimes called interval variables). This type of variable theoretically has an infinite number of values between any two points. Of course in practice the accuracy of measurement will not be perfect, as it will be limited by the observer and the equipment used. Therefore there will only be a limited number of possible values between any two points. Obvious examples of continuous variables are lengths, weights and areas.

> *Note: two words that are often confused:*
> **Accuracy** is the closeness to the real value. This is usually set by the observer or the equipment and should be chosen as appropriate to the variable. When you write down a value it should reflect the accuracy with which the measurement was taken. If you measure to the nearest 0.1 g then 5 g should be written as 5.0 g, not 5.00 g.
> **Precision** is the closeness of repeated measures to the same value. It is possible to have data that are very precise but very inaccurate. For example your balance gives exactly the same value for repeated measures of the same object but they are all overweight because the balance was not calibrated properly. The data obtained would be precise but inaccurate.

Discrete variables (also called 'discontinuous' or occasionally 'meristic'). Unlike continuous variables this type of variable has a limited number of possible values. These possibilities are often, but not always, integers. For example, the number of live-born offspring in a litter of mice can only ever be an integer as there is no possibility of recording a fraction of an offspring.

Discrete variables are often produced by questionnaires. Respondents are offered choices such as: 1, strongly disagree; 2, slightly disagree; 3, neutral; 4, slightly agree; 5, strongly agree. There is clearly a continuous variable ('agreement') here and division of responses into categories in this way is rather arbitrary. It would be very easy to devise different ways of dividing the responses to obtain more or fewer possibilities.

The distinction between discrete and continuous variables can be rather blurred.
• Example 1: a discrete variable becomes continuous. If you measure the number of cells in 1 ml of blood this must be an integer and therefore discrete. However it has so many possible values that it is effectively continuous.
• Example 2: a continuous variable becomes discrete. Seed diameter is a continuous variable but if you measure poppy seed diameter to the nearest 0.05 mm there will be only a few possible values, making it effectively discrete.

How accurate do I need to be? It is often possible to use better equipment or become more careful when measuring to increase accuracy. However, increased accuracy will take longer and will result in fewer data being collected: another trade-off. As a rule of thumb there should usually be between 30 and 300 possible intervals between the smallest and the largest value. If possible, adjust the accuracy of the measurement accordingly. Do not assume that measuring to as many decimal places as possible will make the data any better.

Ranked variables

These occur when data are ordered by magnitude although exact values are not relevant. It is not assumed that the difference between 1 and 2 is the same as that between 3 and 4. Often it is possible to put observations into rank order without measuring at all. For example, plants from six pots could be ranked in 'health' order by simple observation and assigned values from 1 to 6.

Attributes (also called 'categorical' or 'nominal' variables). These variables have few categories — usually 'yes' or 'no', 'male' and 'female' or a small number of possibilities. For example, you could score flower colours as red, blue or yellow. Attributes or categories should, ideally, not have any obvious sequence.

Derived variables (or 'computed' variables). These are usually calculated from two (or more) other variables, e.g. ratios, percentages, indices or rates.

> *Warning: you lose accuracy by combining variables into ratios.*
> For example, if we round to 0.1 then 1.2 implies 1.15–1.25, giving a maximum error of 4.2% and 1.8 implies 1.75–1.85 with a maximum error of 2.8%.
> If these two observations are combined into a single observation then 1.2/1.8 implies a range from 1.15/1.85 to 1.25/1.75 or 0.622–0.714 giving a maximum error of 7%: much greater than the error in the original data.

Distributions of combined variables are often awkward. Be very careful with percentage data as percentages will often have rectangular ('flat' or 'uniform') distributions and/or have limits at 0% and 100%. However, percentages are a familiar and widely used method for expressing observations and there are a few statistical tricks available to help you deal with them.

Ratios can also lead to a loss of information. For example both 5/10 and 500/1000 will give a ratio of 0.5 losing all information about the size of the sample.

Types of distributions

Why do you need to know about distributions?

Just as there are different types of variable, there are different types of distribution. All parametric statistics and many non-parametric ones are based on features of distributions or on assumptions about data following certain distributions.

Discrete distributions

Poisson distribution

This is a very useful tool to use as a starting point in many biological investigations. It is a distribution describing the number of times an event occurs in a unit of time or space. Usually a sample of time or space is taken and the number of events recorded. Typical events are number of fish lice on a fish or number of influenza cases reported in a week.

> *Assumptions of the Poisson distribution*:
> **1** Mean number of occurrences is small relative to the maximum possible.
> **2** Occurrences of one event must be independent of others.
> **3** Occurrences are random.

The purpose of fitting to this distribution is to test for randomness or independence in either space or time. If the number of scale insects on leaves fits a Poisson then it can be assumed that the assumptions hold. Therefore the occurrence of individual insects is unaffected by the presence of others, so we can infer that the scale insects arrive at random and that no leaf is 'full' of scale. If the distribution is significantly different from a Poisson then not all the assumptions hold and further investigation should follow.

Poisson distributions only require knowledge of the mean, as the mean and variance are equal. This property is also very useful as simple inspection of the mean and variance of observations in a sample will give you some idea of the form of a distribution.

If the variance is greater than the mean then the population is more clumped (aggregated) than random. If the variance is less than the mean then it is more ordered (uniform) than random (Fig. 5.1). Distributions may be described by simply quoting their variance/mean ratio — with a value of 1 indicating a random (Poisson) distribution, and higher values indicating clumping.

Binomial distribution

This is a discrete distribution of number of events when there are two possible outcomes for each event and the probability of each is constant. For example, if the probability of each birth producing a female is 0.5 (usually termed p) then the probability of a male is 1 minus 0.5 (also 0.5 in this case and often termed q) as there are no other possibilities. This means that each individual being born has a 50% chance of being female and a 50% chance of being male.

If this is expanded to families with more than one offspring then we can start to apply probabilities to the proportions of males (M) and females (F). For example in a family with two offspring there are four possible outcomes: FF, FM, MF and MM (note that there are two routes to get one male and one female). As the chance of each event has already been determined as 0.5 then the chance of each of the four outcomes is 0.5×0.5 or 0.25. In other words there is a 25% chance of getting FF, 25% for MM and then 25% for each of MF and FM. So 50% of families with two offspring will have one of each sex.

This can be expanded further to three offspring where there are four possible families (reached through eight different sequences, each with a 0.125 probability of occurrence):

Female offspring	Male offspring	Probability	Sequences
3	0	0.125 (1/8)	FFF
2	1	0.375 (3/8)	MFF, FMF, FFM
1	2	0.375 (3/8)	MMF, MFM, FMM
0	3	0.125 (1/8)	MMM

There are many uses of this expansion from single events to groups in biological investigation. To stay with the male–female example for the moment, an

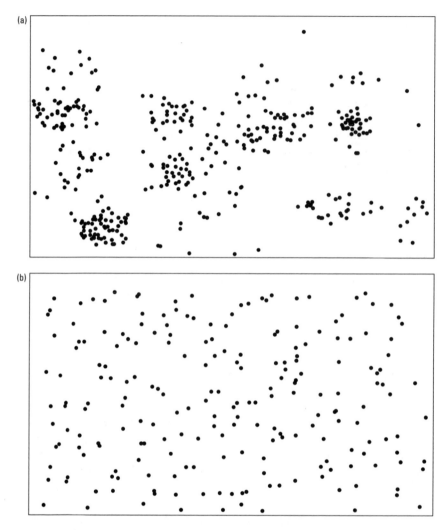

Fig. 5.1 Two hypothetical distributions of individuals in space. (a) The individuals are highly clumped or aggregated. If quadrats were used to sample from this population the variance in number of individuals per quadrat would exceed the mean. (b) In the second distribution the individuals are more ordered than random and the results of number of individuals per quadrat would show a variance less than the mean.

investigation into 480 broods of song thrushes (*Turdus philomelos*) where there were five eggs surviving to fledgings gave frequencies (numbers of observations) for each of the six possible categories of families:

Females	Males	Probability	Expected no.	Observed no.
5	0	0.03125	15	21
4	1	0.15625	75	76
3	2	0.31250	150	138
2	3	0.31250	150	142
1	4	0.15625	75	80
0	5	0.03125	15	23

The expected frequencies from the assumption of a binomial distribution can be tested against the observed numbers using a chi-square test or a G-test. In this case, despite having fewer broods with three of one sex and two of the other than was expected, the difference is not significant and therefore we accept the null hypothesis that the sexes of individuals in song thrush broods of five follow a binomial distribution with a P of 0.5 (i.e. there is a 50% chance of having a female offspring).

The binomial makes a very good starting place for a null hypothesis of even chances of events happening in all groups observed. If the binomial distribution is not followed then alternative explanations about aggregated or dispersed events have to be invoked.

Negative binomial distribution

In many organisms aggregation of individuals in time and/or space is almost ubiquitous. The negative binomial distribution is a discrete distribution that can be used to describe clumped data (i.e. when there are more very crowded and more sparse observations than a Poisson distribution with the same mean). There are reasonable assumptions that can be made about the way organisms distribute themselves that result in a negative binomial distribution. This allows a sensible null hypothesis about aggregated distributions to be made.

Hypergeometric distribution

This is another theoretical, discrete distribution that has some use in biology. The hypergeometric distribution is used to describe events where individuals are removed from a population and not replaced. It is therefore quite useful in small, closed populations that are being sampled destructively and also in the application of mark–recapture techniques.

Continuous distributions

Rectangular distribution (also called a 'flat', 'even' or 'uniform' distribution)

This describes any distribution where all values are equally likely to occur. This distribution rarely appears in reality but it can sometimes be useful for generating a null hypothesis (see Chapter 4).

Normal distribution

This is the most important distribution in statistics and it is frequently assumed that data are distributed in this way. Therefore is it often important to determine whether the data set is a good fit to a normal distribution or not. Methods you can use to test this are the Kolmogorov–Smirnov test, the Anderson–Darling test or a chi-square goodness of fit (see pp. 77, 79 and 68, respectively). These methods are not very sensitive when samples are small and should not be used if there are fewer than about 50 observations.

The normal distribution is a symmetrical, continuous distribution and is described by two parameters — the mean, μ (mu, describing the position), and the standard deviation, σ (sigma, describing the spread). These two parameters are estimated from samples and assigned the letters, m and s.

The normal is sometimes called the Gaussian distribution. A normal distribution always has a characteristic bell shape.

In a perfect normal distribution (where μ is the mean and σ is the standard deviation):

$\mu \pm \sigma$ contains 68.25% of the observations: 50% fall between $\mu \pm 0.674\sigma$
$\mu \pm 2\sigma$ contains 95.45% of the observations: 95% fall between $\mu \pm 1.96\sigma$ (Fig. 5.2)
$\mu \pm 3\sigma$ contains 99.73% of the observations: 99% fall between $\mu \pm 2.576\sigma$

Standardized normal distribution. This is a derived distribution where each observation in a normal distribution is processed by subtracting the mean and dividing by the standard deviation. This gives a normal distribution with a mean of 0 and a variance of 1. One purpose of this transformation is to compare distributions that might have very different means on the same scale to look at the shapes of the distributions.

Convergence of a Poisson distribution to a normal distribution. Even though a Poisson distribution is discrete (you can only get integers), when the mean number of observations is very large a Poisson distribution will approximate to a normal distribution. This could arise, for example, if you counted the number of springtails in a group of soil samples and found that they fitted to both Poisson and normal.

Note: a binomial distribution with more than 100 observations (or fewer if $P \approx 0.5$) will also approximate to a normal distribution.

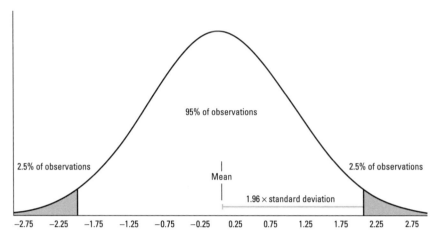

Fig. 5.2 In a normal distribution 95% of the observations will fall within 1.96 standard deviations of the mean. This leaves 2.5% of the observations in each of the tails of this symmetrical distribution (shaded).

Sampling distributions and the 'central limit theorem'. The means of samples taken from any shape of parent distribution will themselves have a normal distribution. This is the basis for the rule that the standard deviation of the sample mean (i.e. standard error) of a sample is $s/\sqrt{n}$, where s is the standard deviation of the observations and n is the number of observations.

Describing normal distribution further

Two types of departure from normality in a data set are skewness and kurtosis.

Skewness. This is another word for asymmetry; skewness means that one tail of the bell-shaped curve is drawn out more than the other (Figs 5.3 and 5.4). Skews are either to the right or left depending on whether the right or left tails are drawn out (i.e. long right tail = right-skewed distribution). Statisticians label the true skewness parameter γ_1 (gamma$_1$) and the estimated value g_1. A negative g_1 indicates skewness to the left and a positive g_1 skewness to the right.

> If a distribution is skewed, the mean is nearer to the tail than the mode and the median, as shown in Fig. 5.3.

Kurtosis. This is a measure of the 'flatness' of a distribution. A symmetrical distribution can differ from the normal in being either **leptokurtic** or **platykurtic**. A leptokurtic distribution has more observations very close to the mean and in the tails. A platykurtic distribution has more observations in the 'shoulders' and

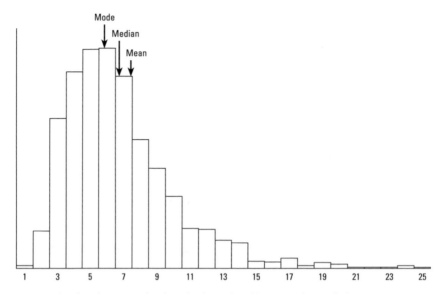

Fig. 5.3 This distribution is clearly right skewed and has a g_1 value well above 0. In a skewed distribution the mean is always nearer the tail than the mode with the median falling between mean and median.

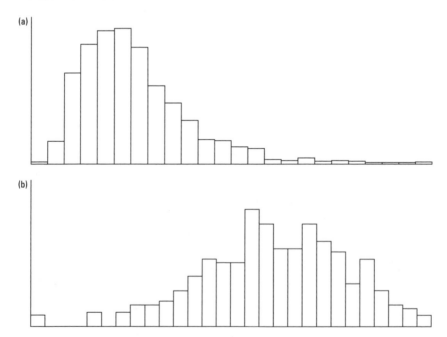

Fig. 5.4 These two frequency distributions are clearly not symmetrical. (a) The data is right skewed and has a g_1 value of 1.53. (b) The data is left skewed and has a g_1 value of –0.335.

fewer around the mean and tails. A bimodal distribution is, therefore, extremely platykurtic. The kurtosis parameter is γ_2 (gamma$_2$) and is estimated by g_2.

In a perfect normal distribution both g_1 and g_2 are equal to 0. A negative g_2 indicates a platykurtic distribution and a positive g_2 leptokurtosis.

Is a distribution normal? It is extremely unlikely that you will collect a data set that is perfectly normally distributed. What you need to know is whether the data set differs *significantly* from a normal distribution. One good way for checking data for departures from 'normality' is to use the **Kolmogorov–Smirnov (K–S) test** or the **Anderson–Darling (A–D) test**. Both tests compare two continuous distributions with the null hypothesis that they are the same (i.e. it tests the sample data against a normal distribution with the same mean and variance as the sample). It is more powerful than the chi-square goodness of fit method which is another commonly used method of determining whether a data set is normally distributed. See Chapter 7 for details of the K–S, A–D and chi-square goodness of fit tests (pp. 77, 79 and 68, respectively).

Transformations

Parametric statistics assume that the data set you are using is distributed normally. So check that this is true first using a statistical test fitting your distribution to a perfect normal distribution with the same mean and variance. If the data set is significantly different from normal try a transformation such as: log, square root, arcsine square root for percentage or proportion data, probits or logits. There are many standard methods to try but as long as you treat each piece of data (datum) in exactly the same way you can do any transformation you like.

For example, a study on feeding preferences in marshland birds counted the number of grey herons (*Ardea cinerea*) seen in creeks and open water at different times of the day. The percentage in creeks is converted to a proportion and the angular transformation and logit transformations given.

Time	No. in creek	No. in open	Percentage in creek	Proportion in creek	Angular transformed	Logit transformed
0600	22	25	46.8	0.47	43.2	−0.13
1200	12	19	38.7	0.39	38.5	−0.46
1800	25	8	75.8	0.76	60.5	1.14

Angular transformation. The **angular** or **arcsine square root transformation** is so routinely applied to percentage data it warrants a description of the method. A percentage is converted to a proportion, this is square rooted and then the arcsin (inverse sin or $\sin^{-1}$) is taken. To make sure you have the calculation correct, either use the values from the table above or check that 10% converts to about 18 after an angular transformation and 100% converts to 90. A common problem encountered with this transformation is that packages use radians rather than degrees and this must be accounted for. Remember that if you are converting direct from percentages rather than proportions, the variable to be converted should be divided by 100 as part of the transformation.

SPSS Assuming the percentages have been converted to proportions and are in a variable called 'prop'. From the 'Transform' menu, select 'Compute'. In the 'Compute variable' box that appears type a name for the target variable (say, 'angular'). Then from the list of functions, select 'ARSIN(numexpr)' and click the up arrow to add it to the 'numeric expression'. Next, with the question mark highlighted in blue, select 'SQRT(numexpr)' from the functions list. Finally, with the question mark selected again, select the variable 'prop' and move it across to the 'numeric expression' (either by double click, or by highlighting and clicking the right arrow). Finally, the correction for converting radians to degrees needs to be applied and the expression multiplied by 57.295.

The 'numeric expression' should now read '57.295*ARSIN(SQRT(prop))'. Click 'OK' and the converted numbers should appear in a new variable.

MINITAB Assuming the percentages have been converted to proportions and are in a variable called 'Prop'. From the 'Calc' menu, select 'Calculator . . .'. In the 'Store results in variable' box, type an appropriate name, such as 'Angular'. Then from the list of 'Functions' highlight 'Degrees' and click 'Select'. Then highlight 'Arcsine' and click 'Select'. 'DEGREES(ASIN(number))' should appear in the 'Expression:' box. Then scroll down the 'Functions' list to 'Square root' and click 'Select'. The 'Expression' should now be 'DEGREES(ASIN (SQRT(number)))'. Double click on 'Prop' from the list on the left and the 'Expression' becomes 'DEGREES(ASIN(SQRT("Prop")))'. Click 'OK' to run the transformation.

(Or, if you have the 'commands' enabled, you could type 'Let "Angular" = DEGREES(ASIN(SQRT("Prop")))' *at the MTB > prompt.)*

Excel Assuming the proportion is in cell A1 the conversion is achieved with the formula: '= DEGREES(ASIN(SQRT(A1)))'. DEGREES, ASIN and SQRT can either be typed in directly or selected from 'Paste function' (f_x) under the 'Math&Trig' submenu. The most common error in calculating the angular transformation in Excel comes from the conversion of radians into degrees.

Logit transformation. Logits are needed for logistic regression. The advantage of a logit transformation is that it converts proportional data limited to 0 and 1 to an unlimited scale by using the likelihood of events. The transformation stretches out values that are near 0 and 1. The logit of a proportion, p, is the natural log (ln or $\log_e$) of p/q where q is the proportion that is not p (i.e. $p + q = 1$).

Note that logits for 0 and 1 are infinite, so will probably give odd results. The logit for 50% should be 0. All values below 50% will be negative, and all above 50% will be positive.

SPSS Assuming the data have been converted to proportions are in a variable called 'prop'. From the 'Transform' menu, select 'Compute'. In the 'Compute variable' box that appears type a name for the target variable (say, 'logit'). Then either, from the list of functions, select 'LN(numexpr)' and click the up arrow to add it to the 'numeric expression', or, type directly into the 'Numeric expression' box. Replace the question mark with 'prop/(1 – prop)', replacing 'prop' with the name of your variable. Click 'OK' and the logit-transformed numbers should appear in a new variable.

If converting direct from percentages, the variable to be converted should be divided by 100 as part of the transformation.

MINITAB Assuming the data have been converted to proportions are in column C1. From the 'Calc' menu, select 'Calculator . . .'. In the 'Store results in variable' box, type an appropriate name, such as 'Logit'. From the list of 'Functions' highlight 'Natural log' and click 'Select'. 'LOGE(number)' will appear in the 'Expression:' box. Replace the text 'number' with 'C1/(1 – C1)', then click 'OK'.

(Or, if you have the 'commands' enabled, you could type 'Let "Logit" = LOGE(c1/(1 – C1)' *at the MTB > prompt.)*

Excel Assuming the proportion is in cell A2 the conversion is achieved with the formula: '= LN(A2/(1 – A2))'. LN can either be typed directly or selected from the 'Paste function' (f_x) under the 'Math&Trig' submenu.

t-distribution

This symmetrical, continuous distribution is related to the normal distribution but is flatter with extended tails. It is the distribution of deviations from the mean divided by the sample standard error of a huge number of samples. As the sample standard error varies between samples the spread is greater than if the deviations were divided by the true standard deviation of the mean (standard error). *t*-distributions have degrees of freedom associated with them that correspond to the size of the sample. So the smallest degrees of freedom of 1 from just two observations will give a very flat distribution and when the degrees of freedom are infinite (i.e. in a sample with an infinite number of observations) the *t*-distribution will recapture the normal distribution.

Confidence intervals

95% confidence intervals are calculated for samples using *t*-distributions (although when the true σ, standard deviation, is known, or the sample size is huge, the normal distribution can be used). 95% confidence intervals should be preferred to the more usually quoted mean ± SE as the standard error of the mean is only really useful if the sample size is known, when it can be converted to a confidence interval of the required width.

> Be very careful when you see headings such as 'Means and standard deviations' as this wording is slightly ambiguous. It usually translates as mean and standard deviation of the observations but is, on occasion, referring to the mean and its standard deviation (i.e. standard error).

Chi-square distribution

This is another continuous distribution that is very useful in statistics. Unlike the normal and the t-distributions it is asymmetric and varies from 0 to positive infinity. The chi-square distribution is related to variance.

> X^2 is the usual way of expressing sample statistics approximating to χ^2.

Exponential distribution

This is a continuous distribution that is occasionally useful as a null model in biology. It occurs when there is a constant probability of birth, death, increase or decrease. So, for example, if a population of beetles invades a new area they may have an exponential increase in numbers as their rate of increase is constant. As soon as the population stops following the exponential distribution the rate of increase has clearly changed. This may indicate that intraspecific competition has reduced the growth rate or a predator is starting to have an effect.

Exponential distributions can also be used to examine decreasing observations. For example, the amount of drug in the bloodstream after an injection may have an exponential decay with 10% being removed every hour. You can test an observed distribution against an expected exponential distribution using a variety of tests of difference.

Non-parametric 'distributions'

It is sometimes better to ignore distributions totally. This is the case when the data set is known to be awkward or difficult to transform. The advantages of making no assumptions about the distribution of the data are considerable as it allows greater flexibility, but there are some limitations in the type of statistical tests that can be used and in the power of the tests.

Ranking, quartiles and the interquartile range

In non-parametric tests data sets are usually ranked before they can be examined statistically (computer packages do this for you). If a set of data is put in rank order from the smallest value to the largest then information about the position of the data set or the spread can be gained by inspecting values at certain points in the ranked data set (for example the median is the value of the data point or datum in the middle of the ranked set).

A quartile is simply the value of the data point that lies a quarter of the way into a data set and it is commonly used to describe the spread of a non-parametric distribution.

The interquartile range is the difference in the values between the data point one-quarter of the way down the ranked list to the point three-quarters of the way down.

Box and whisker plots (a.k.a. box plots)

These summarize data where there are no assumptions of distribution. A sample is represented by a box whose top and bottom represent the upper and lower quartiles (i.e. the box covers the interquartile range). The box is divided at the median value. A line (the whisker) is drawn from the top of the box to the largest value within 1.5 interquartile ranges of the top and the same from the bottom. Any values outside this range are then added as symbols (see Fig. 6.1). These outliers are often identified in some way (they certainly are in the statistical package, SPSS) so you can check them. Outliers are the values most likely to have been mistyped!

6

Descriptive and presentational techniques

These techniques are presented in roughly the same order as they appear in the key.

General advice. Descriptive and presentational techniques serve two rather different purposes. The first is to summarize and display data in the best way possible for a reader to derive information about the data. If this is the intention then the techniques used should be as simple as possible and require the minimum effort from the reader. The second purpose is for researchers to explore their own data. A variety of methods should be employed that show the data from different perspectives. In this way you can become familiar with your data and may be stimulated to pursue new lines of enquiries or test different hypotheses.

This chapter is intended to offer general advice on data presentation and although all the examples are generated in the statistical packages featured in the next chapters there are no detailed descriptions for navigating the menus to generate the figures you see.

Displaying data: summarizing a single variable

Box and whisker plot (box plot). This is an excellent way of summarizing data, especially if it is not normally distributed. The plot shows the median value as a thick bar, the interquartile range as a box and the full range as the 'whiskers'. Some statistics and graph drawing packages show outliers (data points well outside the range of others) as individuals points. An example is shown in Fig. 6.1.

Displaying data: showing the distribution of a single variable

It is important that any graphical depiction of data is clear. Usually the easiest and clearest way to display a single set of data is to use a histogram or a bar chart of frequency of occurrence. If you have discrete data then it may be best to display each possible value. However, in most cases it will be necessary to group the data into classes. There is often confusion about the difference between a histogram and a bar chart.

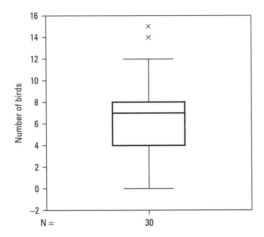

Fig. 6.1 This box and whisker plot was created in SPSS. There were 30 observations of numbers of bird species seen by a single observer from the same point during a fixed time. It shows that the median number was seven and that 50% of the observations were between 4 and 8. Note that there were single observations of 14 and 15 (marked as crosses) and that the axis extends to –2 even though 0 is clearly a lower limit (it is impossible to see fewer than zero species of birds).

Bar chart (for **discrete data**). Each possibility is represented on the abscissa (*x*-axis), and frequency on the ordinate (*y*-axis). Gaps between the bars symbolize the discrete nature of the data (Fig. 6.2). If there are a very large number of possibilities then a bar chart may be inappropriate as clumping the data into groups will give a better picture of the distribution. If this happens then you have moved to a histogram.

Histogram (for **continuous data**). Observations are grouped into artificial classes. The mid-point of the class is displayed as a label on the *x*-axis and frequency (number of observations) on the *y*-axis. No gaps should be left between classes to symbolize the continuous nature of the data. Shading, especially intense shading, should be used sparingly (Fig. 6.3).

Number of classes to display in a histogram?

As a rule of thumb use 12–20 classes (categories along the *x*-axis)...
    But employ some common sense. Small samples should rarely need to have 12 classes and huge samples may be grouped into more than 20 classes.
    As an alternative rule of thumb use $\sqrt{n}$ classes (where *n* is the number of observations in your sample).
    In the example of the beetle elytra (Fig. 6.3) there are 19 classes of 1.5 mm each although only 15 have any observations. This fits with the first rule of thumb. The second rule of thumb suggests 24 classes.

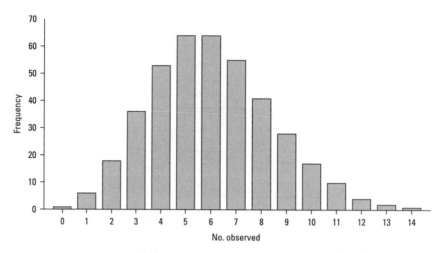

Fig. 6.2 This bar chart was generated using a larger version of the same data set as was used to create the box and whisker plot in Fig. 6.1. This SPSS chart shows the number of bird species seen in a garden in a 15 min period. There were 400 observations made. Clearly the number of birds can never be below 0 although it might be greater than 14. Observations of this kind will always be integers although there is certainly no requirement for data to be integers to be suitable for bar charts. Gaps between the bars symbolize the discontinuous nature of the data.

Fig. 6.3 This histogram presents a data set comprising 589 observations of elytra length in a population of beetles. The observations are linear measures and clearly continuous. All measurements were made to the nearest 0.1 mm. Each bar represents a range of values and there are no gaps between the bars. Values on the *x*-axis show the mid-points of the range for half the bars.

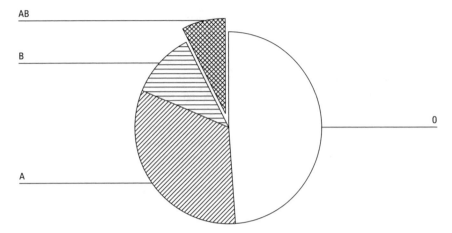

Fig. 6.4 This pie chart, generated in SPSS, shows the blood groups of a sample of 200 people. A pie chart is appropriate for this sort of data as if it was presented as a bar chart the *x*-axis would have no real meaning. Shading is not required but may be used if desired. Slices may be 'exploded' for emphasis, as the AB slice has been in this example.

Pie chart (for **categorical data** or **attribute data**). A pie chart should only be used if the categories have no logical sequence. For example if the categories are blood groups, species of tree or mutants of *Drosophila* then a pie chart is probably a better method of presentation than a bar chart. However, if the categories have a logical sequence, such as five arbitrarily defined levels of ripeness, then a bar chart will be more informative. An example is shown in Fig. 6.4.

Tip: do not use 3D bars or shadow effects on histograms or bar charts (unless it is for a display and then only in exceptional circumstances). Such effects obscure the data as it is difficult to see exactly where the top of the bar lies. I would also advise against the use of colour unless it is absolutely necessary (although I cannot think of an example where I would advocate its use!).

Descriptive statistics

Statistics of location

There are several ways of defining the 'location' of a distribution. It is tempting to focus only on the arithmetic mean as this is the easiest statistic to calculate and the most commonly used. However it is worth considering some of the alternatives, especially the median.

Arithmetic mean. This is the 'normal' mean, often called an average and by far the most commonly used measure of location; when written it is usually denoted as x bar (i.e. $\bar{x}$), which is an estimate of the true mean which is represented by the Greek letter, μ (mu) sometimes written as μ_x.

Geometric mean. This is the antilog of the mean of the logged data; it is always smaller than the arithmetic mean. The most commonly encountered use of this statistic is when data have been logged or when data sets that are known to be right skewed are being compared.

Harmonic mean. This is the reciprocal of the mean of the reciprocals and is always smaller than the geometric mean. This type of mean is rarely needed.

Median. This is the middle value of a ranked data set. After the arithmetic mean it is the next most commonly used measure of location. It is the measure highlighted in box and whisker plots. If all the data are put into rank order (arranged in a list from the largest value to the smallest) the median is the value associated with the middle ranked item (half-way down the list).

Mode. This is the most 'fashionable' value in a set of data — the value that occurs most frequently. It can be used with any type of data, even categorical.

- Variables with one clear mode are said to be unimodal.
- Distributions with two peaks are bimodal.
- Distributions with more than two peaks are multimodal.
- The trough between two modes is sometimes called the antimode (Fig. 6.5).

One of the problems with the use of the mode is that it is rarely suitable if the observations are made with any degree of precision (e.g. femur length to the nearest 0.01 mm) as there will be a much lower chance of an observation being repeated. Therefore the mode should only be used when there are either a very large number of observations or a fairly small number of possible values.

Note: in any unimodal, symmetrical distribution (for example, a perfect normal distribution) the mean, median and mode are all the same (see Fig. 5.3 to see what happens in an asymmetrical distribution).

Statistics of distribution, dispersion or spread

There are several ways to display the distribution or spread of a set of observations. However, it is important the measure used is appropriate to the data *and* the statistic of location (e.g. median) used.

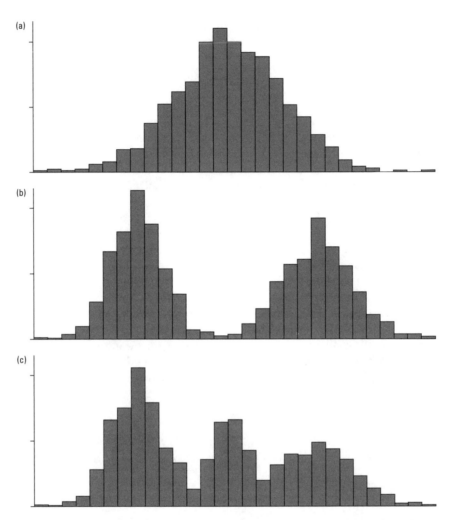

Fig. 6.5 Three rather different frequency distributions. (a) There is a clear single mode of a unimodal distribution. (b) There are two almost totally distinct distributions giving a bimodal distribution. This might indicate two separate populations, different genders or different species. (c) The pattern of the frequency distribution is even more complex. There are three distinct modes making a multimodal distribution. This may indicate, for example, three cohorts of recruitment into a population.

Range. This is the most basic measure of dispersion and is simply the difference between the largest and smallest observations in a sample. It is usually quoted as the smallest and largest value (e.g. range = 9.76–15.23 cm).

Interquartile range. This is a non-parametric measure of dispersion that works on the ranked data. It is the difference between the value of the data item (datum) 25% of the way down a ranked list and the one 75% down. These values are called quartiles. The interquartile range is much more useful than the range as it is unaffected by outliers. Unlike many other measures of dispersion, the interquartile range is not necessarily symmetrical about the median. The quartiles are often given the codes 'Q1' and 'Q3'.

Variance. The variance usually refers to the sample variance, s^2, which is an estimate of the true variance, σ^2 (sigma squared). It is the mean of the squared deviations of observations from their arithmetic mean. Variance is rarely used as a descriptive statistic as it is not in the same units as the original observations. However, many statistical tests use variance in their calculations.

Standard deviation (SD). This is usually an estimate, s, of the true standard deviation, σ (sigma). It is the square root of variance. This is commonly used as a descriptive statistic as it is in the same units as the original measurements or observations. However, confidence intervals should be used if comparisons of different sets of observations are required.

Standard error (SE). By convention this is short for 'standard error of the mean' (i.e. the standard deviation of a distribution of means for repeated samples from a population). Standard errors are often quoted with means although this is probably because they are small rather than for any good statistical reason! If several samples are to be compared then the confidence interval should be preferred. If a measure of the variation in the sample is required then standard deviation is better.

In theory there is no difference in calculation between a standard error and a standard deviation, it is just that the former measures the standard deviation of a hypothetical sample of means.

Confidence intervals (CI) or confidence limits. These are derived from the standard error of the mean. Confidence intervals are the most useful measure of the dispersion of a distribution.

If a sample from a population is very large then the true mean of the population is 95% likely to lie within 1.96 standard errors of the sample mean. This region is called the 95% confidence interval as you are 95% certain that it contains the true mean of a population. As samples get smaller then the multiplier used gets larger and the confidence intervals get wider. (If you have statistical tables it is easy to determine the required multiplier as it is derived from the t-distribution.) Confidence intervals are always symmetrical about the arithmetic mean. They are to be preferred over standard errors if several sets of observations are being compared.

Coefficient of variation. This is used to compare the amount of variation in populations with different means where direct comparisons of the standard deviations (s) are difficult to make as they are confounded by differences in scale. The coefficient of variation is usually denoted V or CV. $CV = (100s)/$mean and is usually expressed as a percentage.

Other summary statistics

There are other components of the shape of the distribution of observations that can be interpreted easily. Knowledge of the skewness of a data set is particularly useful.

Skewness. This is a measure of the symmetry of a data set. If the data set is symmetrical then the value of skewness will be 0. If there is a tail to the right it will be positive; if there is a tail to the left it will be negative. Meaningful values for skewness are only possible if there are more than 30 (and preferably a lot more) observations in the data set. Normal distributions are symmetrical and consequently have a skewness of 0. Skewness is discussed in Chapter 5.

Kurtosis. This is a measure of the shape of a distribution. It tells you whether there are more observations around the mean or less when compared to a normal distribution. Meaningful values are only possible if there are more than 100 observations in the data set.

Using the computer packages

All statistical packages will give summary statistics for sets of observations. However, generating exactly the set of statistics you are interested in may take several steps. The less frequently used statistics such as kurtosis may not be available.

SPSS In this package the data may appear to be in the same spreadsheet form as a package such as Excel but the approach is rather different as the statistics are not displayed on the spreadsheet but in a separate window. The data should be in a single column with an appropriate label. To change the column label simply double-click on the column name ('var0001' by default), or click on the 'Variable View' tab, and replace the 'Name' with something more suitable (you are limited to eight characters and spaces are not allowed). The screen shot shows the various measures of dispersion that are available under the 'Descriptives...' options in the 'Descriptive Statistics' submenu of the 'Analyze' menu. The default selections are shown, which include the rarely useful minimum and maximum (Fig. 6.6). Once you have chosen the options you want the statistics

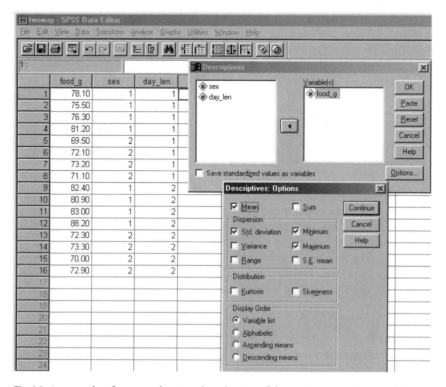

Fig. 6.6 A screenshot from SPSS showing the selection of descriptive statistics available.

displayed in the 'Output' window. As you proceed through an SPSS session output accumulates in this window. This is very useful as you can go back and check results of previous tests very easily.

Further descriptive statistics can be accessed. From the 'Analyze' menu select 'Descriptive statistics' then 'Frequencies...' selection. Select the 'Statistics...' button in the dialogue box and an array of options such as skewness, kurtosis, variance, mode and median are available.

MINITAB The data for a single variable should be in one column in the spreadsheet section of the package. The variable should be named appropriately in the cell under 'C1'. Spaces are allowed as part of the label. To get simple descriptive statistics go to the 'Stat' menu and select 'Basic statistics' and then 'Display Descriptive statistics...'. Move the name of the column with the data from the list on the left into the 'Variables:' box. Either click 'OK' now or take a detour to the 'Graphs...' options first to request some graphical output.

This is the sort of output that is generated:

Descriptive Statistics: Height

Variable	N	Mean	Median	TrMean	StDev	SE Mean
Height	24	11.257	11.294	11.253	0.198	0.040

Variable	Minimum	Maximum	Q1	Q3
Height	10.838	11.776	11.129	11.383

All the basics are reported here. The number of observations in the data set ('N'), the arithmetic mean ('Mean'), largest and smallest values ('Maximum' and 'Minimum'), standard deviation ('StDev') and standard error ('SE Mean') as well as the non-parametric versions of these statistics: 'Median' and the upper and lower quartiles ('Q3' and 'Q1'). Quartiles are explained further elsewhere in this chapter. There is also the rather unorthodox trimmed mean 'TrMean' where the tails of the distribution, the top and bottom 5% in this case, are removed before a mean is calculated. This makes the estimate of the mean less likely to be affected by outliers.

If the 'Graphical summary' is selected as an option in the 'Graphs...' options box, there is considerably more output to assess (Fig. 6.7).

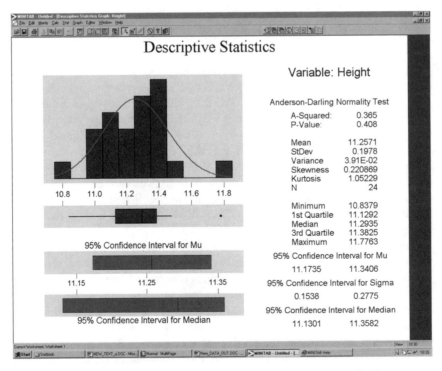

Fig. 6.7 A screenshot from MINITAB when the 'Graphical summary' option has been chosen to display descriptive statistics.

This output contains much of the same information as the non-graphical version but with some extras. In the mass of output on the right, the first thing is a test for normality. This is a test to determine whether the data in question are normally distributed. The 'A-Squared' value is the output from a test and 'P-Value' is the probability that the data are normally distributed. If the *P*-value is less than 0.05 then this means it is unlikely to be normally distributed and therefore parametric statistics should not be used.

After this test comes the more usual descriptive statistics of arithmetic mean, standard deviation, variance and then the measures of the 'shape' of the distribution: skewness and kurtosis and the number of observations '*n* of data'. Next comes some information about the data arranged in rank order. The value of the smallest and largest observations and then observations one-quarter (first quartile), half (median) and three-quarters (third quartile) of the way down a ranked data list.

The last section gives 95% confidence intervals for three of the descriptive statistics: 'Mu' (the arithmetic mean), 'Sigma' (the standard deviation) and the median.

On the left of the output are four graphs. First is a histogram of the raw data with a normal distribution superimposed on it (the normal distribution shown has the same mean and standard deviation as the data). Then comes a box and whisker plot of the data (described elsewhere in this chapter) and finally graphical representations of the mean ('Mu') and median with their 95% confidence intervals.

Excel In this package you have to assign a cell of the spreadsheet to contain the summary statistic you require. Assign a cell by clicking on any empty cell. Then you identify the cells that contain the variable (raw data) that you are interested in and the statistic appears. For example: your data, containing 100 observations, has been typed into the first column of the spreadsheet (column A). The first cell has the title of the variable and the actual observations are in rows 2–101. To calculate the arithmetic mean of this variable you go to any cell and declare its contents as '=AVERAGE(A2:A101)'. (As you can see Excel calls the arithmetic mean the 'average'.) The median can be calculated as '=MEDIAN (A2:A101)', the geometric mean by '=GEOMEAN(A2:A101)' and the harmonic mean as '=HARMEAN(A2:A101)'.

These and other summary statistics are easily accessed using the 'Paste function' (f_x) facility of the package, mostly in the 'Statistical' submenu. Or, once you have learned a few of the function codes you could just type them in. For instance '=STDEV(A2:A101)' to get the standard deviation reported in an empty cell. Most of the summary statistics mentioned in this chapter are readily available in Excel, including skewness ('Skew'), kurtosis ('Kurt'), mode ('Mode'). Confidence intervals require more work, but are possible using 'Confidence'.

The interquartile range is not given directly in Excel, but it is readily calculated. '=QUARTILE(A2:A101,1)' will give the value of the first quartile (i.e. datum 25% up the data set when sorted) and '=QUARTILE(A2:A101,3)' will give the third quartile (75%). The interquartile range is the difference between these two values, so in Excel that would be: '=QUARTILE(A2:A101,3) − QUARTILE (A2:A101,1)'.

Displaying data: summarizing two or more variables

Box and whisker (box plots). These are a good way of comparing two variables. They allow direct visual comparison of both the location and the dispersion of the data. An example of the use of two box plots is shown in Fig. 6.8.

Error bars and confidence intervals. A similar way of looking at the same data is to display the arithmetic mean and some measure of the dispersion of the data. An example of the use of mean and confidence interval is given in Fig. 6.9. Note that the interquartile range is not symmetrical about the median (Fig. 6.8) whereas the 95% confidence intervals (or standard deviation if we had chosen to display that instead) are symmetrical about the mean (Fig. 6.9).

You can display more than two groups using these methods. They provide a very powerful method of showing differences and similarities between many groups. In the example here there is almost no need for any further statistics — the difference between males and females is so striking!

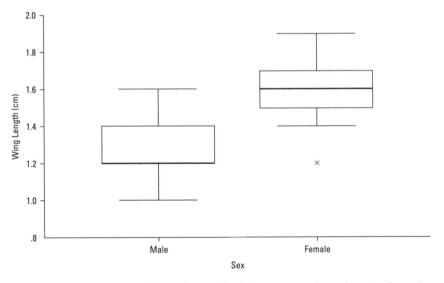

Fig. 6.8 In this spss-generated figure the sample of observations of wing lengths of a moth are divided into two groups by gender. As in most insects, the females are considerably larger than the males and although there is some overlap in the 'whiskers' there is no overlap of the interquartile range of the two groups. Note that for the males the median and lower quartile are superimposed showing that 25% of the observations for males were almost of the same value.

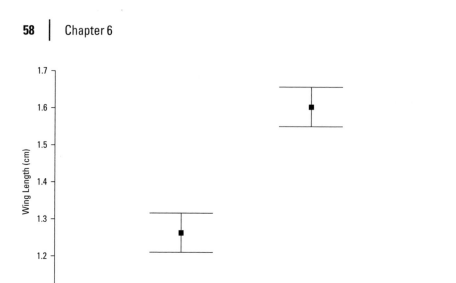

Fig. 6.9 This figure, generated in SPSS, uses the same moth wing lengths as Fig. 6.8. The means for males and females are represented by filled squares and the 'whiskers' are error bars that extend to cover the 95% confidence interval for the mean (i.e. there is a 95% chance that the true mean of the population lies between the extremes shown). There is no overlap between the 'whiskers' suggesting that the groups are likely to be highly significantly different.

Displaying data: comparing two variables

Associations

These can be used if two observations are made from a single individual (e.g. the 'individual' is a stream, and the water pH and stream flow have been recorded). Before any statistics are applied it is best to get a 'feel' for the observations by a graphical representation of the data.

Scatterplots. The simplest way to display a relationship between two variables is to use a plain scatterplot (Fig. 6.10). This assumes that two observations on the same row in the package are two measurements from the same 'individual'. An 'individual' can be almost anything: sampling station, greenhouse, pair or single bone.

It is important that all figures should have appropriate axis labels on them. They should also be accompanied by a figure legend that makes the plot interpretable without reading the relevant section of the text.

Do not add extra information that is not relevant or appropriate. For example many packages offer best-fit lines as a simple option. Do not use these unless: (i) you believe there is a 'cause' and 'effect' relationship between the variables; (ii)

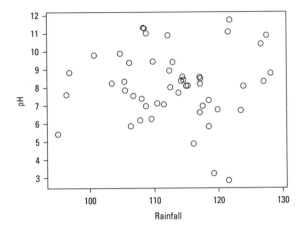

Fig. 6.10 The scatterplot of pH and rainfall from a range of sites shown here has been created in MINITAB using the default options. Clearly there is no obvious relationship between the two variables.

you have used regression and you want a graphical accompaniment; (iii) you intend to use regression; or (iv) you wish to use one variable to predict the other.

Multiple scatterplots. A good way to compare observations from two sites where the same variables have been recorded is to use a multiple scatterplot. The axes will be exactly the same as for the single scatterplot but each group will be displayed using a different symbol. This technique works particularly well for two or three groups and less well for more than that. Choose symbols carefully to make the groups easily distinguished and make sure that the figure caption makes it clear which symbol matches which group (Fig. 6.11).

A more sophisticated use of symbols can convey a great deal of information about several factors on the same scatterplot. For example, if the data for two morphological variables are collected and the individuals are divided into groups by sex and species then all this information can be incorporated in a single plot. This can be achieved by using shaded and non-shaded symbols for the two sexes and different shapes for the two species (see Fig. 9.1 for an example).

Trends, predictions and time series

Lines. These should only be used to join points if there is a reasonable assumption that observations could be made between the points (Fig. 6.12). This is perfectly reasonable if the x-axis is temperature with readings made at 15–35°C in steps of five as intermediate temperatures are valid. However if the x-axis is number of eggs in a nest and the y-axis egg weight then it is perhaps unwise to draw a line linking mean weight at four eggs with mean weight at five as the line will pass though impossible points.

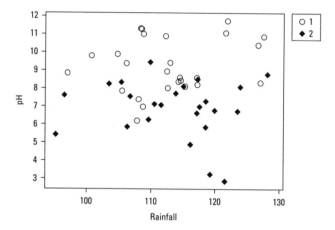

Fig. 6.11 In the example shown here, two sets of observations from different study areas are identified with different symbols. A quick glance shows that group 1 is associated with a higher pH than group 2 but there is no obvious difference between the groups on the 'rainfall' axis. An analysis of variance or *t*-test could be used to determine the statistical probabilities, but the results would only confirm what is obvious from the scatterplot.

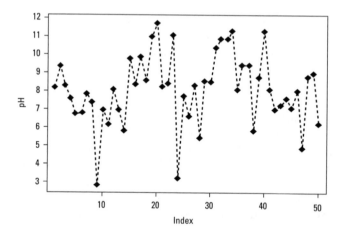

Fig. 6.12 This MINITAB-generated example of a line graph shows a set of 50 readings of pH made through time at a chemical plant. The time gaps were equal and the observer thought it valid to join the reading made with lines, as there is a reasonable expectation that the intervening times would have intermediate levels of pH.

If there are several observations for each point on the *x*-axis then it is usually better to plot the mean or median with a measure of the dispersion at each point rather than use a scatter of points (Fig. 6.13). The same guidelines for joining means apply as for joining single observations.

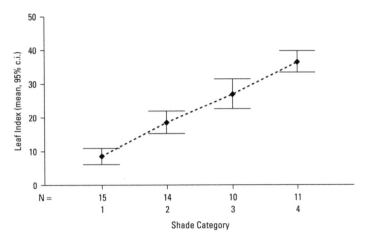

Fig. 6.13 This figure, generated in SPSS, shows a combination of a line graph and the mean and confidence interval approach of Fig. 6.9. In this case there are four levels of variable shade that can be said to form a valid sequence from light to dark. Observations of leaf shape (a continuous variable) were taken at each of the four shade categories and the mean and 95% confidence intervals (c.i.) are plotted here with the means joined to emphasize the clear trend.

It is very easy to deceive a reader by altering scales. For example, if there is a slight but steady increase in the concentration of nitrate in a lake over time then this can be made to look like a rapid increase if the scale on the y-axis starts not from 0 but from a value just below the lowest observed value. This kind of manipulation of the reader will work particularly well if there is no measure of the variation given.

Fitted lines. The best way to draw the reader's eye to a relationship between two variables is to use a fitted line of some sort. Indeed, an observer can sometimes be fooled into seeing a relationship in a scatterplot when there is none (Fig. 6.14). For this reason the use of fitted lines should be restricted to circumstances when the line is meaningful. The most common use is to illustrate a relationship between two variables that has been investigated using regression.

One technique is to plot the scattered observations along with the fitted line and then give more information about the regression in the text or the figure legend.

Confidence intervals. These should always be used to show the reliability of a mean value, as shown in Figs 6.9 and 6.13. If a regression line has been calculated then it should always be displayed with its confidence intervals. This shows the range within which the line is 95% likely to lie (Fig. 6.15). If the confidence intervals are wide apart then the line is obviously less reliable.

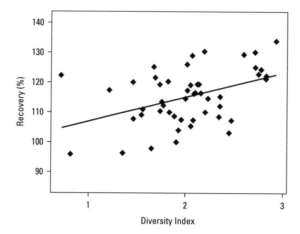

Fig. 6.14 This MINITAB-generated scatterplot of the recovery of biomass after an extreme event against the diversity index before the event shows a slight but non-significant trend. However, the addition of the trend line draws the eye and emphasizes the slight trend, convincing the reader that the trend is real.

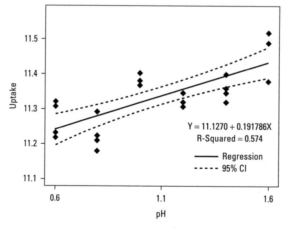

Fig. 6.15 Here the raw data, regression line and the 95% confidence intervals of the regression line are all shown, along with some of the regression output from MINITAB. The variable 'uptake' measures the amount of drug passed across the stomach lining of a rabbit at various experimental pH levels. There is a clear relationship: the regression line slope is significantly different from 0 and it explains 57.4% of the variation in the uptake observations. The 95% confidence intervals (CI) are quite close to the best-fit line confirming that the relationship is robust. Note that the line, quite properly, does not extend beyond the data as predictions can only be safely made within the range of the data.

Displaying data: comparing more than two variables

Associations

It may be tempting to use the full capacity of the graphics on the package you are using but there is little or nothing to be gained by plotting a multidimensional plot that is impressive to look at but impossible to interpret.

Three-dimensional scatterplots. This type of figure looks impressive but is quite difficult to interpret for several reasons associated with representing three dimensions in a 2D medium:

1 If there are too many points plotted then those nearest the 'front' will obscure those as the back.

2 As the display medium is 2D all the points need to have 'spikes' to anchor them to the '$x = 0$, $z = 0$' plane. Without these spikes then a point near to the front but high on the y-axis will look *identical* to one near the back but low on the y-axis. An alternative method of spiking is to attach all points to the origin. This occasionally is useful, but normally generates a figure that looks like a bunch of flowers.

3 There is often no forced perspective making the arrangement of the axes seem odd. Furthermore points at the back are usually the same size as those at the front and this fools the eye (Fig. 6.16).

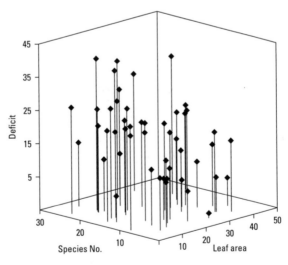

Fig. 6.16 Most statistical packages have a 3D plotting ability (this one is plotted using SPSS). The 3D scatterplot is very difficult to interpret, even when the relationship is quite strong and impossible when the relationship is weak. The spikes make the figure very cluttered but are vital to place the point accurately.

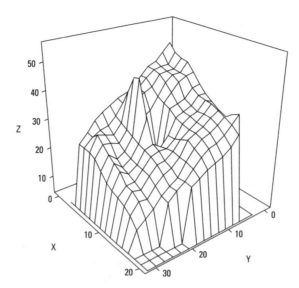

Fig. 6.17 Possibly even more difficult to interpret than the 3D scatterplot shown in Fig. 6.16 is the 3D surface plot, such as this one drawn in MINITAB. This sort of figure can only be interpreted if relationships are very strong or the smoothing algorithm is so strong that all the variation is wiped out of the data. Particular problems of 3D surfaces are that the edges tend to be extrapolated from far fewer data points than the middle and that peaks can obscure a lot of the surface behind them.

Multiple trends, time series and predictions

Multiple fitted lines. Further information may be conveyed if two lines are fitted on the same graph. The advantage of this approach is that lines may be compared directly but the disadvantage is that the message may become confused. I would advise against a tactic I have seen used increasingly, which is to have different *y*-axes for the same *x*-axis so that the two lines being compared fit sensibly. There are two problems with this type of graph. First it makes the reader see relationships that are not really there and second it is often difficult to see which scale applies to which line.

Surfaces. Many statistical packages include the option to have spectacular 3D surface plots. I would advise against the use of these in almost all situations. The problems of all '3D' graphs on 2D surfaces apply with the additional problem that the solid or apparently solid surface totally obscures much of the surface. The way the points are connected to form the surface is questionable too; for example, if one or two of the axes have data that are normally distributed. This means that these observations are contributing a great deal of information in the centre of their range and rather less at the extremes. The surface plot does not reflect this in most cases (except it often betrays this by tending to have

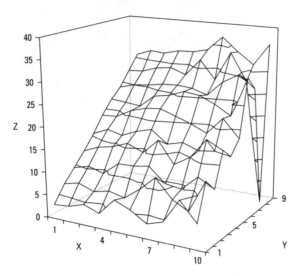

Fig. 6.18 This 3D scatterplot from Excel shows a common problem with interpolation of points. The surface is constructed from an array of 100 values arranged in a 10×10 grid. Unfortunately one value is missing at $x = 10$, $y = 9$ and this value has been interpreted as having a value of 0.

smoother edges where the surface is extrapolated from fewer points). Therefore the edges of the surface can be influenced by the extreme points, the very points likely to be measured with less accuracy.

Remember that a surface, like a joined up line, can only be used if both the 'x' and 'y' (or 'z') observations can reasonably be expected to have possible intermediate values (Fig. 6.17). A problem with the interpolation of missing data points is shown in Fig. 6.18.

7

The tests 1: tests to look at differences

These are presented in roughly the order they appear in the key.

Do frequency distributions differ?

There are two basic types of questions that can be asked:
1 Does one observed set of frequencies differ from another?
2 Do the observed frequencies conform to a standard distribution?

In the first case the test becomes an analogue of a two-sample test of difference, such as the *t*-test. In the second it is a way of testing observations against expected frequencies, such as in plant-breeding crosses when particular ratios of phenotypes are expected or to test if organisms are occurring at random by testing against the Poisson distribution. The **G-test**, **chi-square goodness of fit**, **Kolmogorov–Smirnov** and **Anderson–Darling tests** are the most commonly employed tests to answer these questions and are described below.

G-test

In situations where you have observed frequencies of various categories and expected proportions for those categories that were not derived from the data themselves then the G-test should be the preferred statistic to use. However, it is only a few years since this test was shown to be superior to the traditional chi-square goodness of fit approach on theoretical grounds. Consequently it is not supported by any of the packages considered in this book. If your package supports the G-test then use it and its associated correction factor, Williams correction.

In the G-test the ratio of the observed and expected frequencies is calculated. The natural log (ln or $\log_e$) of this ratio is calculated and these values are multiplied by the number observed, summed, then doubled. This value of G is then compared to a chi-square distribution with one fewer degrees of freedom than the number of categories.

Example. A dihybrid cross of sweet peas has four categories of plant types with an expected ratio of $9 : 3 : 3 : 1$. This ratio was not generated by the data so when the data are collected it should be compared to the expected values using the G-test. Two hundred plants were collected. The number of plants in each of the four categories is given below.

Tall and pink	Tall and white	Dwarf and pink	Dwarf and white
108	35	46	11

SPSS This test is not available in SPSS.

MINITAB This test is not available in MINITAB.

Excel As this is a relatively simple calculation it is ideal for a spreadsheet like Excel. The most likely source of error when carrying out a G-test is to miscalculate the expected frequencies. The total of the expected frequencies should be exactly the same as the number observed.

1 Label five columns as: observed, expected ratios, expected frequencies, ratio and ln ratio.

2 Enter the four observed values in cells A2 to A5. (Your data may have more than four categories and therefore all references to row five here should be adjusted to your data set.)

3 Enter the values 9, 3, 3, 1 in cells B2 to B5.

4 In cell A7 use '= sum(A2 : A5)' to sum the number of observations.

5 Copy cell A7 across to B7: '= sum(B2 : B5)'.

6 In cell C2 the expected frequency of tall and pink plants is needed. This should be 9/16 of the total number of observations. Use '= B2/B$7*$A$7', which will give 112.5 in this example, and copy this cell down to C5.

7 Calculate the ratios in D2 as '= A2/C2'. Copy this cell down to cell D5.

8 Calculate the natural logs and multiply by the number of observations (A2) in E2 as '= A2*ln(D2)'. Copy down to cell E5.

9 Sum the values in column E in cell E7 and double it: '= 2*sum(E2 : E5)'. The example should give a value of 2.33. If your data gives a negative value adjust the contents of cell E7 to '= −2*sum(E2 : E5)' and the value should become positive.

10 The value in E7 should be looked up on a chi-square table. Excel does this with '= CHIDIST(E7,3)' (or select CHIDIST from the 'Paste function', Statistical submenu). Three is the degrees of freedom in this example. The result is $P = 0.505611$, so the null hypothesis is not rejected (Fig. 7.1). Your calculation should have one degree of freedom fewer than the number of categories.

11 Save your Excel sheet as it will be easy to recycle for future G-tests by altering the values in columns A and B (where necessary). If more categories are required remember to adjust all the summing steps to cover the required ranges and adjust the degrees of freedom.

Fig. 7.1 Calculating the value of G using Excel.

Chi-square test (χ^2)

Often known as the chi-square goodness of fit, this test is one of the most widely used in the whole of biology. It is also the statistical test you are most likely to be familiar with. You will usually present the data in a table showing the observed and expected frequencies for various categories. These categories can be single outcomes or groups of possible outcomes. It is customary to use grouping of categories to ensure that none of the expected values is less than 1 (some authors, erring on the side of caution, suggest 5). The expected values can either be derived from a distribution such as the Poisson or negative binomial, they can assume that all categories are equally likely (a flat or rectangular distribution), they can be derived from a specific null hypothesis of ratios or they can be derived from another set of data. In all cases the null hypothesis (H_0) will be that the observed and expected frequencies are not different from each other. The chi-square test may also be used as a test of association (see Chapter 8).

Example. A very common starting point for investigations in biology is to determine whether events or observations are occurring at random. If events are truly random then they should follow a Poisson distribution (see Chapter 5 for more details). The chi-square test allows you to compare observed data with the expected data if following a Poisson distribution with the same mean. In this case the number of lice found on adult char is recorded. All observations were taken from a single catch of 98 fish. If the lice attach themselves to the fish at random then they will follow a Poisson distribution and the chi-square will not be significant (i.e. the null hypothesis is that lice attack fish at random). If the result is significant this indicates that lice do not attack randomly and a new hypothesis should be formulated.

Note that in this instance we are comparing a set of observed frequencies against a set of expected frequencies derived from a Poisson distribution that has the same mean as the observations. Therefore the expected frequencies are not independent of the observed and this loses us a degree of freedom.

No. lice/fish	0	1	2	3	4	5	6	7	8+
No. observations	37	32	16	9	2	0	1	1	0

SPSS First I should point out that processing this type of data for a goodness of fit chi-square in SPSS is not easy unless you wish to fit the distribution to a uniform one (i.e. all categories are expected to have the same number of observations). If you have this sort of data use the Kolmogorov–Smirnov test to answer the question in SPSS. However, I will go through the procedure anyway assuming you do not know how to calculate Poisson 'expecteds' by hand (perhaps the faint hearted should move on to the K–S test now!).

1 Make sure that the data are in a single column with the actual data, not the frequencies. In this case there will be 98 rows in the data set: one for each fish. Label this column 'no_lice'.

2 Determine the mean number of lice/fish using the 'Analyze' menu, then 'Descriptive statistics...' and selecting either 'Descriptives...' or 'Frequencies ...' before moving 'No_lice' into the 'Variable(s)' box. The output will confirm the mean as 1.1429.

The next steps show how to calculate the expected frequencies in SPSS. If you can do this by hand or in a spreadsheet I suggest you do so and skip directly to step 8.

3 Generate a new column with all the expected frequencies in it (e.g. 0, 1, ... 7) in separate rows. Label this column 'freq'.

4 Use 'Compute...' which is under the 'Transform' menu to bring up a dialogue box. In the 'Target variable' box enter, say, 'exp1'. Then scroll down the list of functions until you reach 'CDF.POISSON(q,mean)' (CDF stands for cumulative density function). Select this and move it into the 'Numeric expression' box with the up arrow. Replace the second question mark with the mean you calculated in step 2 (i.e. 1.1429 in the example). The first question mark ('q') should be replaced with the name of the variable you created in step 3 (i.e. 'freq'). Click 'OK'. A new variable will appear on the spreadsheet with numbers starting from 0.32. This first number is the probability of getting a 0 in a Poisson distribution with a mean of 1.143.

5 Unfortunately the other values are cumulative values so the number in the second row is the probability of getting a 0 or a 1. The actual probability of getting a 1 can be calculated by hand using simple subtraction (i.e. the probability of getting a 1 is the probability of getting 0 or 1 minus the probability of getting a 0). The probability of getting exactly 2 is the difference between the cumulative value for 1 and 2. When your calculated probabilities drop below 0.01 (because with a sample of 98 this will drop you below the critical threshold of an expected value of 1 or more which is the limit for a chi-square test) move to step 8. If you have a high mean it may be easier to do another couple of steps in SPSS . . .

6 Repeat step 4 but use a different 'Target variable', say 'exp2' and replace 'q' with 'freq-1' instead of just 'freq' (assuming you labelled the list of numbers as 'freq'). This will produce a new column of probabilities with a 0 at the top.

7 One final calculation step is required. Go to the 'Transform' menu and select 'Compute. . .' again. Insert a new label in the 'Target variable', say 'expected' as this is going to be the true expected value. Select 'exp1' from the list on the left and move it to the 'Numeric expression' box. Press '−' and then move 'exp2' over too. Put the expression 'exp1 − exp2' in brackets then add '* 98' (or whatever your sample size is if you are not using the example) outside the brackets. This will multiply the values by the total number of observations, turning your expected frequencies into expected numbers of lice.

8 You will notice that all expected frequencies for more than 4 lice/fish are less than 1 and should be grouped together to form a category of four and above. Write down the expected frequencies for the five categories.

No. lice/fish	0	1	2	3	4+
No. observations	37	32	16	9	4
Expected	31.3	35.7	20.4	7.8	2.8

9 As all values above 3 lice/fish have been grouped for expected frequencies, this must now be done for the actual data. In SPSS you can either do it by hand or use the 'Recode' option under the 'Transform' menu. Selecting either an 'Into same variables. . .' (overwriting the original data) or 'Into different variables. . .'. If you choose the latter, move 'no_lice' into the 'Numeric variable → output variable' box. Then type a name for your new variable in the 'Output variable, Name' box and click 'Change'. Your new name appears after the old name in the main box. Then select 'Old & New Values'. This brings up a bewildering set of options that are actually extremely useful in SPSS. In the 'Old value' section on the left select the button for 'Range' with a box and the word 'through highest' under it. Put a 4 in the box. Then in the 'New value' section on the right put a 4 in the 'value' box and click 'add'. This will put '4 thru highest → 4' in the 'old → new' section (meaning that all fours or higher will be fours). Finally click 'All other values' on the left, 'copy old values' on the right, and 'add' to keep the rest as they were. 'ELSE → Copy' appears in the 'old → new' box. Click 'Continue' here and then 'OK' in the next window to create the new variable.

10 Finally, choose 'Analyze', 'Nonparametric tests' and 'Chi-square...' to bring up the 'Chi square test' window. Move the variable you created in step 9 into the 'Test variable list:' box. Unfortunately you have to enter the expected frequencies one at a time. Click on values in the 'expected values' area and enter the expected frequencies, starting at the one for zero (i.e. 31.3 in the example) and clicking 'add' after each one before clicking 'OK' to run the test. This is the output you should expect.

NPar Tests

Chi-Square Test

Frequencies

NAME

	Observed N	Expected N	Residual
.00	37	31.3	5.7
1.00	32	35.7	-3.7
2.00	16	20.4	-4.4
3.00	9	7.8	1.2
4.00	4	2.8	1.2
Total	98		

Test Statistics

	NAME
Chi-Square [a]	3.069
df	4
Asymp. Sig.	.546

a. 1 cells (20.0%) have expected frequencies less than 5. The minimum expected cell frequency is 2.8.

The data have been reduced to just five categories and each has an associated expected frequency. The residual is the difference between observed and expected. At the bottom is the actual chi-square statistic, the associated degrees of freedom (4 in this case as, once the categories had been clumped, there were five categories) and finally the significance value (P-value).

Note: as the mean of the distribution we are comparing to our data was calculated from the sample we should lose a further degree of freedom to give a total degrees of freedom of 3. The package does not account for this as it is calculated before the test is applied and therefore SPSS 'knows' nothing about it.

Despite the fact that there are more high values and 0 values than you would expect (indicative of a clumped distribution) the probability is well above 0.05. This indicates that the deviation from the Poisson expectations is non-significant and therefore we have no reason to reject the null hypothesis that lice attack fish at random in this population.

MINITAB Calculating a goodness of fit to a Poisson distribution is surprisingly awkward in MINITAB, although not quite as awkward as in SPSS. However, it is a test that you will wish to carry out in many circumstances.

1 Make sure that the data are in a single column with the actual data, not the frequencies. In this case there will be 98 rows in the data set — one for each fish. Label the column 'No lice'.

2 Determine the mean number of lice per fish using the 'Stat' menu, then 'Basic statistics', then 'Display Descriptive statistics'. Move 'No lice' into the 'Variables' box. The output will confirm that there are 98 observations and give the mean as 1.143.

(*Or, if the command interface is enabled, type* 'Describe "No lice"' *at the MTB > prompt.*)

The next steps show how to calculate the expected frequencies for a Poisson distribution in MINITAB. If you know how to do this by hand or in a spreadsheet you can skip directly to step 6.

3 First you should generate a table of the tallied observations: Go to the 'Stat' menu, then 'Tables' then 'Tally...'. Move 'No lice' into the 'Variables' box and make sure that the 'Counts' box is checked. Click 'OK'.

This output will appear in the 'Session' window:

```
Tally for Discrete Variables: No lice

No lice   Count
     0      37
     1      32
     2      16
     3       9
     4       2
     6       1
     7       1
    N=      98
```

Note that as there were no fish with five lice, there is no count for five in the table. Now you can either cut and paste the two columns of figures from the

Fig. 7.2 Generating expected values using the 'Tally' command and Poisson probabilities in MINITAB.

'Session' window into columns C2 and C3 of your MINITAB spreadsheet (remembering to click on the 'Use spaces as delimiters' option) or you can type the numbers in directly.

(*Or type* 'Tally c1;' *at the MTB > prompt, followed by* 'Store c2 c3.' *at the SUBC > prompt. Remember to type a semicolon at the end of MTB > command if you want to bring up the SUBC > prompt.*)

4 Now we need to generate the expected numbers of lice for a Poisson distribution with the same mean as the sample. Go to the 'Calc' menu, then 'Probability distributions' then 'Poisson...'. In the dialogue box type the sample mean in the 'Mean:' box, type 'c2' in the 'Input column:' box and 'c4' in the 'Optional storage:' box. If you do not select a storage column the output only goes to the 'Session' window (Fig. 7.2). Click 'OK'.

(*Or type* 'PDF c2 c4;' *at the MTB > prompt, followed by* 'Poisson 1.143.' *at the SUBC > prompt. Replace 1.143 with the mean of your sample.*)

5 To convert the probabilities generated into numbers you need to multiply by the total number of observations in the sample (98 in this example). Go to the 'Calc' menu, then 'Calculator...'. In the dialogue box type c5 in the 'Store result in variable:' box. Then type 'c4 * 98' in the 'Expression:' box (replacing 98 with the number of observations in your sample). Click 'OK'.

(*Or type* 'Let c5 = c4 * 98' *at the MTB > prompt.*)

6 Chi-square tests should not have expected frequencies that are less than 1. In the example the expected frequencies for six and seven are both less than 1. Also, because there were no fish with five lice in the example, there is no expected frequency for five in the column. In this example the best strategy is to pool all observations of 4 or more into a single observed and expected value. You can do this in several ways, either on a calculator, by hand or by using the 'Calculator...' from the 'Calc' menu, selecting a row and typing the sum required — in this example selecting row 5 and typing '98 − 31.248 − 35.717 − 20.412 − 7.777' in the 'Expression box'. The numbers are the expected frequencies of 0, 1, 2 and 3 lice per fish, respectively. The result should be 2.85, being the expected number of times, out of 98, that a Poisson distribution with a mean of 1.143 would produce a number of 4 or more.

(*Or type* 'Let c6(5) = 98 − 31.248 − 35.717 − 20.412 − 7.777' *at the MTB > prompt. Replace c6 and (5) with the column and row you require and the numbers with those appropriate for your data.*)

7 You must amalgamate the observed frequencies in exactly the same way as the expected frequencies. In this case there are a total of four observations of four or more lice per fish. You should now have one column of expected frequencies with no values of less than 1 and one of observed. In the lice and fish example the following columns should result:

C6 Expected	C7 Observed
31.2484	37
35.7169	32
20.4122	16
7.7771	9
2.8460	4

8 Finally we reach the chi-square test itself. Amazingly there is no way to reach the required test using the menus. You will have to make sure the command line is enabled (activate the session window, then from the 'Editor' menu select 'Enable commands'). Type at the command line (MTB >) the following: 'LET K1 = SUM((C7 − C6)**2/C6)' (assuming that your expected values are in C6 and observed in C7) followed by 'PRINT K1' to see the result. For the example this will return the value of '3.059' (slightly different numbers will be the result of rounding in the calculations).

9 Finally, to determine whether this is significant or not (i.e. do we reject the null hypothesis that the lice are attacking the fish at random), we need to be sure about the degrees of freedom. In the example there were five possible values (so 4 d.f.), but the mean value of the Poisson distribution we are comparing against was taken from the data we are using so we lose a further degree (leaving 3 d.f.).

You can either look up the value on a table in a statistics book or you can use the command line in MINITAB using the following commands:

```
MTB > cdf k1 k2;
SUBC > chisquare 3.  (replacing 3 with the d.f. required)

MTB > let k3 = 1 - k2
MTB > print k3
```

This will give you the output:

```
Data Display
K3        0.382556
```

This value is well above the critical 0.05 level. Therefore we have no reason to reject the null hypothesis that lice attack fish at random in this population. If you look up the value in a statistics book you will see that the chi-square value required to reach the critical level of 0.05 for 3 d.f. is above 7 and here the chi-square value was only just above 3.

Excel Most of the calculation steps are fairly straightforward and therefore ideal for a spreadsheet like Excel. Using the same example as for SPSS (above) I will assume that the data are available as a frequency table rather than a column containing all 98 observations. If it is not you can generate a frequency table from the raw data using the 'Frequency' command (or by hand).
1 To calculate Poisson expected values the only parameter you need to know is the mean. To find the mean of data in a frequency table you first need to find the product (what you get when you multiply the category value by the number of observations). If you have input the categories in column A with a title, and the number of observations in B with a title, then go to cell C2 and type '= a2*b2'. In this example this will be 0. Then find all the other products by dragging the little square in the bottom right corner of the cell down to the bottom of the list. Add up the number of observations in the B column by typing 'SUM(b2 : b9)' in cell B11 and dragging this across to the C column to add up the products. Then divide the products by the number of observations to obtain the mean by typing '= c11/b11' in cell B13. Always use labels in other cells to make everything clear.
2 Calculation of the expected number of observations is surprisingly easy in Excel. Move to cell D2 (for convenience) and click on 'Paste function'. Select 'Statistical' and then 'POISSON' from the very long list of options you are offered, click 'OK'. A window with three lines to fill appears. In the first 'X' line use the cell number of the category, in this case A2. In the second type in the mean (1.143); do not use the cell number. Type a 0 in the bottom line. This tells Excel that you do not want the cumulative probabilities. Then click on 'OK'. The probability of getting a 0 in a Poisson distribution with a mean of 1.143 appears in the selected cell (just under 0.32).

Fig. 7.3 Calculating the mean number of lice per fish from a table of frequencies in Excel. Note that the numbers and letters around the sides of the data are the Excel row and column labels.

3 Step 2 calculated the probability of getting a particular number of lice. What is needed is the expected *number* of observations. To get this you must multiply the probability by the number of observations, 98 in this example. To do this make sure that cell D2 is selected and add a '*98' to the end of the formula giving: '= POISSON(A2,1.143,0)*98'. Or, to make the calculation more versatile, use the cell where the total number of observations is calculated (b11 in the example, so b11). Press return and the number in the cell becomes the expected number of observations. Remember to replace 1.143 with your mean and 98 with the number of observations in your sample or use cell numbers.

4 Select D2 again and drag its contents down the column (click on the little square in the bottom right corner of the cell and hold the mouse button as you move the mouse to highlight all cells to D9 then let go). The column fills with the expected number of observations for each of the categories in column A. Label column D 'expected' (Fig. 7.3).

5 Some of the expected values are less than 1 and need to be amalgamated. In this case it is best to have a category for lice numbers of four and above. Copy the observed data into column E but replace the number of times four lice were observed with the number of times four or more lice were observed. This is a total of four observations in this case. Label column E 'observed'.

6 You have to amalgamate the expected frequencies too so that cell D6 holds not the expected number of fours but the expected number of 'four or mores'. There are several ways of achieving this but probably the easiest is to start with the 98 total and take off the expected number of 0, 1, 2 and 3. Type in cell D6 '= 98 – D2 – D3 – D4 – D5'. This should give you an expected value of 2.845.

7 Chi-square uses the formula: '(observed – expected)^2/expected' for each category and then sums these to produce the final chi-square statistic. The formula is often quoted as $(O – E)^2/E$. In this example the observed values should be in column E and the expected values in column D. So in cell F2 type in the formula: '= (E2 – D2)^2/D2'. Once you have done that copy it down the F column to F6.

8 You should have five numbers in column F. These need to be added up to give the final chi-square value. Type in any clear cell: '= sum(F2 : F6)'. For convenience I used cell F8. You should get the value: 3.0599. This is the chi-square value that you would quote in a report.

9 Finally, what is the probability of getting this value of chi-square (or a higher value)? Go to any clear cell. Click on 'Paste function' and select 'CHIDIST' from the list (if it does not appear immediately it can be selected from either the 'Statistical' list or the 'All' list). In the first, 'X', box you input the chi-square value (or the cell that you used in step 8). In the second box you need to put the degrees of freedom. In this case there were five categories, giving 4 d.f., but the Poisson distribution we are comparing against has its mean taken from the frequency data which loses another degree of freedom. Therefore you should have 3 d.f. Input 3 and you should get a probability of 0.38. This value is well above 0.05 so you can infer that the distribution of lice on fish is not significantly different from random (Poisson distribution). We have no reason to reject the null hypothesis that lice attack fish at random in this population.

Kolmogorov–Smirnov test

The Kolmogorov–Smirnov test for goodness of fit has a variety of uses for large samples of continuous data. There are two main forms called the one-sample test or two-sample test. Both are used to compare two sets of data to determine whether they come from the same distribution. The one-sample version is more commonly used and compares experimental data with expected distributions. The expected distribution may be derived from the data or may be completely independent of them. For example, you may use the test to determine whether a set of tarsus length data differs from a normal distribution with the same mean and variance as the sample data before you use parametric analysis on it. The two-sample test can be used to compare a set of egg weight data from a population of ducks with a set from another site asking whether the distributions are the same.

Note: although the K–S test appears similar to the *t*-test and the Mann–Whitney *U*test, it is not aimed at the same question. The K–S test delivers a probability that two distributions are the same while the *t*-test is concerned with means and the Mann–Whitney *U*test with medians. Two distributions may have identical means and medians and yet have differences elsewhere in their distributions.

Example. The weight in grams is recorded for a sample of 48 mice. This sample is part of an experiment and the researchers wish to know whether the weights are distributed normally before they go on to use parametric statistics.
  Here are the data:

12.5	13.5	13.2	12.5	12.1	12.6	12.1	12.8
14.2	13.2	13.8	12.0	12.5	12.1	12.8	12.9
12.6	12.8	12.5	13.1	12.4	13.5	13.4	13.6
13.0	14.1	12.6	13.2	13.8	13.8	13.9	14.0
14.1	12.1	12.9	14.5	13.2	14.1	12.5	12.5
15.0	12.6	13.0	13.5	14.0	12.9	12.4	12.8

SPSS This is a very simple test to access in SPSS. Ensure all the data are in a single column. Select the 'Analyze' menu, then 'Nonparametric tests', then '1-sample K–S ...'. In the dialogue box move the name of the column you are testing into the 'Test Variable List:' box. Make sure that the 'Test distribution' has 'Normal' selected. Then click 'OK'. The following output will appear:

NPar Tests

One-Sample Kolmogorov-Smirnov Test

		MOUSE_WT
N		48
Normal Parameters [a,b]	Mean	13.1083
	Std. Deviation	.7178
Most Extreme Differences	Absolute	.115
	Positive	.115
	Negative	-.082
Kolmogorov-Smirnov Z		.795
Asymp. Sig. (2-tailed)		.552

a. Test distribution is Normal.

b. Calculated from data.

This output confirms the test and variable used, and gives statistics: the number of observations ('N'), mean and standard deviation. The last two lines of the output table refer to the test itself. The only important bit is the 'Asymp. Sig. (2-tailed)'. If this number is less than 0.05 then the distribution of the data is significantly different from normal. In this case the value is 0.552 which is well above the critical 0.05 so we have no reason to suppose that the distribution of the data is not significantly different from a normal distribution.

MINITAB The Kolmogorov–Smirnov is a very simple statistic to reach if you are testing a single column to see whether it follows a normal distribution or not, although the Anderson–Darling test, considered below, is easier to reach.

Ensure all the data you wish to test are in a single column. Select the 'Stat' menu, then 'Basic statistics...', and the 'Normality Test...'. In the dialogue box move the column you wish to test into the 'Variable:' box and select the Kolmogorov–Smirnov test. Then click 'OK'. The output is a graph that shows a perfect normal distribution of data as a straight line and your data as a series of dots. The test compares the dots with the line. In the bottom left corner is the output from the statistical test. If the 'Approximate P-value' is less than 0.05 then the distribution is significantly different from normal. In this case the value is given as > 0.15 so it is not significantly different from normal.

The Ryan–Joiner test for normality and the Anderson–Darling test are also offered in this dialogue box.

(Or, *if the command interface is enabled, at the MTB > prompt in the session window type* '%NormPlot C1;' *then at the SUBC > prompt type* 'Kstest.')

Excel There is no direct method.

Anderson–Darling test

One of many tests commonly encountered to test whether a set of data follows a normal distribution or not. The P-value reported is the probability of the data being normally distributed. If $P < 0.05$ the data does not follow a normal distribution and parametric tests should not be used.

SPSS Not available in this package.

MINITAB There are two routes to the Anderson–Darling test in MINITAB. The first is described above for the Kolmorogov–Smirnov test. The Anderson–Darling test is also part of the extensive 'Graphical summary' output for simple variable descriptions. The data should be in a single column. From the 'Stat' menu select 'Basic Statistics' then 'Display Descriptive Statistics...'. Move the name of the column containing the data into the 'Variables:' box. Select the 'Graphs' option and then the 'Graphical summary' option. Click 'OK' twice. An output containing several graphs and many descriptive statistics appears. The first part of the numerical output give the statistics for the Anderson–Darling test ('A-squared') followed by the P-value associated with the statistic. If the P-value is less than 0.05 the data are significantly different from a normal distribution and it is inadvisable to use parametric statistics.

Excel Test of normality are not available in this package.

Do the observations from two groups differ?

The two groups can be paired, repeated or related samples or they can be independent. Paired measures are considered first.

Paired data (a.k.a. related or matched data)

Paired samples or paired comparisons occur when a single individual is tested twice (e.g. before and after) or the sampling station is retested. Another possible use occurs when an individual is, or individuals of a clone are, divided and then subjected to two treatments. Three tests are considered below: the **paired t-test**; **Wilcoxon signed ranks test** and the **sign test**.

Paired t-test

The data must be continuous and, at least approximately, normally distributed. The variances of the two sets must be homogeneous (this can be tested by the Levene test). The null hypothesis is that the there is no difference between the two columns and they could come from the same data set.

Example. It is suggested that the building of a power station will affect the amount of particulate matter in the air. However there are only three readings available for the month before the project got underway. The sites where the three readings were taken were revisited once the station was complete.

Case (site)	Before	After
1	34.6	41.3
2	38.2	39.6
3	37.6	41.0

SPSS Arrange the data into two columns of equal length such that each row represents one individual (or site in this case). The columns should be labelled 'before' and 'after' as this will make it easier to interpret the output.

Under the 'Analyze' menu choose 'Compare means' and then 'Paired-samples t test'. In the dialogue box that appears highlight both 'before' and 'after' as your paired variables then move them across to the 'Paired variables' box with the right arrow. Then click 'OK'.

The output will come in two parts. The first gives some information about the data like this:

Paired Samples Statistics

		Mean	N	Std. Deviation	Std. Error Mean
Pair 1	BEFORE	36.8000	3	1.9287	1.1136
	AFTER	40.6333	3	.9074	.5239

Paired Samples Correlations

		N	Correlation	Sig.
Pair 1	BEFORE & AFTER	3	-.749	.462

It tells you the names of the two variables, the means, how many observations there were, the standard deviation and the standard error of the two sets of data. The next table refers to a Pearson product–moment correlation done on the data with the value of 'r' given as 'Correlation' and 'Sig.' the test of the null hypothesis that 'r' is 0 (i.e. no correlation).

The second part of the output is the report from the t-test itself:

Paired Samples Test

		Paired Differences							
					95% Confidence Interval of the Difference				
		Mean	Std. Deviation	Std. Error Mean	Lower	Upper	t	df	Sig. (2-tailed)
Pair 1	BEFORE - AFTER	-3.8333	2.6764	1.5452	-10.4820	2.8153	-2.481	2	.131

The first three columns refer to the mean difference between pairs of data, followed by the standard deviation and standard error of the differences. Next comes the confidence interval of the mean saying it is 95% likely to be between the two values given. The 95% confidence interval is very wide because there are only three pairs of data. Finally comes the t-test itself with the result of the test (the value given can be looked up on a Student's t-table). Then the degrees of freedom (number of pairs minus one) and the probability of this t-value (or larger) occurring if the null hypothesis is correct. In this case the probability is 0.131 (or 13.1%) and therefore we conclude that there is no significant difference between the two columns. However, with such a small set of data, achieving a significant result is extremely unlikely.

MINITAB First input the data into two columns: one for before, the other after. There is no need to have a separate column to label the individual sites although this might help you interpret the results.

1 From the 'Stat' menu, select 'Basic Statistics' and then 'Paired t . . .'. Move the two variables into the boxes labelled 'First sample:' and 'Second sample:'. Click 'OK'.

You get the following output from the example:

```
Paired T-Test and CI: Before, After

Paired T for Before - After

                    N      Mean     StDev    SE Mean
Before              3     36.80      1.93      1.11
After               3     40.63      0.91      0.52
Difference          3     -3.83      2.68      1.55

95% CI for mean difference: (-10.48, 2.82)
T-Test of mean difference = 0 (vs not = 0): T-Value = -2.48  P-Value = 0.131
```

2 The output confirms that you have run a paired t-test of before and after. There are three pairs of data in the example giving a value of 3 for 'N' with summary statistics. A line on the table of 'Difference' is important as the test is actually comparing the values of difference against zero. Finally a t-value is given and then the important value, the P-value. If this is less than 0.05 then you must reject the null hypothesis that the difference between before and after is 0. In this case the value of 0.131 indicates that there is not enough evidence in the three pairs of data to reject the null hypothesis. However, with such a small data set in the example it is not surprising that you cannot detect a significant effect.

3 The null hypothesis is that the differences are not significantly different from zero. If you want to test a difference other than zero, or you want to have a one-tailed test (i.e. you are only interested in the hypothesis that 'before' is greater than 'after', with the null hypothesis that 'before' is *not* greater than 'after') then go to 'Options' in the 'Paired-t' dialogue before running the test.

Excel The data should be input in two columns of equal length. If the spreadsheets were set up using the example exactly as the table above, then the data for 'before' would be in cells b2, b3 and b4, while that for 'after' would be in cells c2, c3 and c4. In Excel it does not really matter where the data are on the spreadsheet as long as you know the cell numbers.

Select an empty cell where you want the result reported. Then there are two ways of achieving the same result.

1 Go to 'Paste function', select 'Statistical' and then 'TTEST'. Define Array 1 as 'b2 : b4' and Array 2 as 'c2 : c4'. Select 'Tails' as '2' (you will nearly always require a two-tailed test) and the 'type' as '1' (this selects a paired test in Excel). The probability or *P*-value of '0.131' will then appear in the cell.

2 Type in 'TTEST' followed by the first and last cell of the first column separated by a colon, then the same for the second column. Then type the number of tails in the test (usually 2) and then a 1 to ask for a paired test. In this case you would type 'TTEST(b2 : b4,c2 : c4,2,1)' and the probability will appear in the cell.

Wilcoxon signed ranks test

This test is the non-parametric equivalent of the paired *t*-test. It has far fewer assumptions about the shape of the data although it does assume that the data are on a continuous scale of measurement. This means that any type of length, weight, etc. will be suitable. The test is somewhat less powerful than the paired *t*-test. A minimum of six pairs of data are required before the test can be carried out.

Example. In this example the 'individuals' are sampling stations in a river system and the data are measures of flow. The investigator wishes to know if the flow is significantly different on the 2 days. The null hypothesis is that there is no difference in flow.

Station	Day 1	Day 2
1	268	236
2	260	241
3	243	239
4	290	285
5	294	282
6	270	273
7	268	258

SPSS Arrange the data into two columns of equal length such that each row represents one individual. The columns should be labelled. From the 'Analyze' menu choose 'Nonparametric tests' and then '2-related samples'. As a default the 'Wilcoxon' test should be checked but ensure this is the case. Select the two variables ('day_1' and 'day_2' in this example), move them to the 'Test Pair(s) List:' box with the right arrow, then click 'OK'.

The output will look like this:

Wilcoxon Signed Ranks Test

Ranks

		N	Mean Rank	Sum of Ranks
DAY_2 - DAY_1	Negative Ranks	6ª	4.50	27.00
	Positive Ranks	1ᵇ	1.00	1.00
	Ties	0ᶜ		
	Total	7		

a. DAY_2 < DAY_1

b. DAY_2 > DAY_1

c. DAY_1 = DAY_2

Test Statisticsᵇ

	DAY_2 - DAY_1
Z	-2.197ª
Asymp. Sig. (2-tailed)	.028

a. Based on positive ranks.

b. Wilcoxon Signed Ranks Test

The test classifies the paired data into three categories: those where 'day_2' is less than 'day_1' ('Negative Ranks'), those where it is greater ('Positive Ranks') and those where they are the same ('Ties'). In this case there are six of the first, one of the second and no ties. It also ranks the absolute differences (from smallest difference as rank 1 to the largest as rank 7). In this case the one pair where the flow is greater on 'day_2' is also the smallest difference as it is ranked '1.00'. The output value of 'Z' in the second table may be looked up in statistical tables but the result is given anyway as a 'Asymp. Sig. (2-tailed)'. In this case $P < 0.05$ so the null hypothesis must be rejected. The alternative hypothesis that the flows were different on the 2 days is accepted.

MINITAB This test is not achievable in a single step. However, if the data are arranged in two columns you can carry out an analogous test to the paired t-test.
1 You need to create a new column that contains the difference between the first observation and the second. Go to the 'Calc', select 'Calculator...'. Type 'Diff' in the 'Store result in variable:' and 'Day 1'–'Day 2' (either by typing or double-clicking) in the 'Expression:' box (replacing with the names of your variables as appropriate).
2 The null hypothesis is that the median of the differences is not significantly different from zero. This is tested using a one-sample Wilcoxon test. Go to the 'Stat' menu, then 'Nonparametrics', then '1-sample Wilcoxon ...'. In the dialogue box

put 'diff' in the 'Variables:' box. Select the 'Test median' option, leave the value as 0.0 and make sure that the 'not equal' option is selected from the pull-down menu. Click 'OK'.

You get the following output from the example:

Wilcoxon Signed Rank Test: Diff

```
Test of median = 0.000000 versus median not = 0.000000

                  N for   Wilcoxon             Estimated
             N    Test    Statistic       P     Median
Diff         7       7        27.0    0.035      10.50
```

---

3 The output confirms that you are testing the null hypothesis that the median of 'Diff' (i.e. the difference between the flows on the 2 days) is zero (with the package deciding to give a ridiculous number of decimal places). The important value is the *P*-value 'P' which is 0.035 in this example. This is less than the critical 0.05 and the null hypothesis should be rejected. The alternative hypothesis H_1 is that the median of the differences is not equal to zero. This means that the flows on day 1 and day 2 were different. Inspection of the raw data or the estimated median shows that flow was greater on day 1.

Excel There is no easy method.

Sign test

This is a very simple test that makes almost no assumptions about the form of the data, only that it is possible to compare them in some way to decide which is larger. The test is of very low power but is very safe (i.e. it is a conservative test and type I errors are very unlikely). The sign test should only be used when there are large numbers of paired observations.

The test works using the assumption that if two sets of observations are not different then there will be the same number of pairs when A is bigger than B as there are when B is bigger than A. Therefore the actual values of the data points are relatively unimportant as long as they can be compared to see which is the larger. This means the test is not very sensitive to poor quality of data.

The sign test uses the binomial distribution (p. 35) to calculate significance values.

Example. The same data as were used for the Wilcoxon test.

SPSS Arrange the data into two columns of equal length such that each row represents one individual (or site). The columns should be labelled appropriately. From the 'Analyze' menu choose 'Nonparametric tests' and then '2-related samples'. As a default the 'Wilcoxon' test should be checked; uncheck this and check 'Sign' instead. Select the two variables ('day_1' and 'day_2' in this example), move them to the 'Test Pair(s) List:' and click 'OK'.

The output using the example data described earlier for the Wilcoxon signed ranks test will look like this:

Sign Test

Frequencies

		N
DAY_2 - DAY_1	Negative Differences[a]	6
	Positive Differences[b]	1
	Ties[c]	0
	Total	7

a. DAY_2 < DAY_1

b. DAY_2 > DAY_1

c. DAY_1 = DAY_2

Test Statistics[b]

	DAY_2 - DAY_1
Exact Sig. (2-tailed)	.125[a]

a. Binomial distribution used.

b. Sign Test

For the seven pairs of observations there are six where 'day_2' is less than 'day_1' ('Negative differences'). The probability of this or a more extreme result occurring by chance is 0.1250 or 12.5% and this is given as the 'Exact Sig. (2-tailed)'. In fact it is giving the probability of getting 0, 1, 6 or 7 out of 7 in the same direction, given that we are expecting the chance of getting a higher or lower value to be 0.5 (i.e. an equal chance of A being bigger than B as for B being bigger than A).

In this case we would not reject the null hypothesis that the 2 days had different flows. This shows clearly that the sign test is of very much lower power than the Wilcoxon signed ranks test. In fact the sign test is really only useful as the number of paired observations becomes quite large.

MINITAB This test is carried out in a very similar way to the Wilcoxon signed ranks test. Arrange the data in two columns and label the columns appropriately.
1 Create a new column that contains the difference between the first observation and the second. Go to the 'Calc', select 'Calculator...'. Type 'Diff' in the

'Store result in variable:' and 'Day 1'–'Day 2' (either by typing or double-clicking) in the 'Expression:' box (replacing with the names of your variables as appropriate). Obviously, you will have to replace these names with the appropriate names for the data columns.

2 The null hypothesis is that there are equal numbers of positive and negative differences. This is tested using a one-sample sign test. Go to the 'Stat' menu, then 'Nonparametrics', then '1-sample Sign...'. In the dialogue box put 'diff' in the 'Variables:' box. Select the 'Test median' option, leave the value as 0.0 and make sure that the 'not equal' option is selected from the pull-down menu. Click 'OK'.

You get the following output from the example:

```
Sign Test for Median: Diff

Sign test of median = 0.00000 versus  not =  0.00000

              N  Below  Equal  Above      P    Median
Diff          7    1      0      6   0.1250   10.00
```

3 The output confirms that you are testing the null hypothesis that the median of 'diff' (the difference between the flows on the 2 days) is zero. The important value is the P-value 'P' which is 0.1250 in this example (i.e. there is a 12.5% chance of getting this result or a more extreme one). This is greater than the critical level of $P = 0.05$ and the null hypothesis should be accepted. These are the same data as were used in the Wilcoxon signed ranks test, which rejected the null hypothesis, thus demonstrating what a conservative test it is.

Excel There is no direct method. However, if you compare each pair of data and count up cases where the first data set is greater then you can use the 'Binomdist' function in Excel to work out the P-value. The binomial distribution is described in Chapter 5. Here I give two methods for the sign test in Excel and a method for counting up the number of '+' and '−' pairs.

Using the example above, there are seven pairs of data. In six of the seven the first column is greater. We are assuming a null hypothesis that there is no difference between the two columns. If that is the case then we expect an equal number of observations where the first column is greater and when the first is smaller.

1 In Excel the probability of getting six plusses out of seven can be determined fairly easily. After selecting an empty cell where you want the result to be displayed there are two methods. *Either*: use 'Paste Function', select 'Statistical', then 'BINOMDIST', click 'OK' and you will be confronted with four boxes. In the first type the number of successes ('Number_s'), in this case 6. In the second the number of 'Trials', in this case 7. Next the probability of getting a 'success', in this case we choose 0.5 (in other words, we assume an equal chance of getting a plus or a minus). In the last box type 0 to indicate that you do not require the cumulative probability. Click 'OK' and this will give you the result 0.054. *Or:* type in the values described above directly: '= BINOMDIST (6,7,0.5,0)'.

This is not the complete answer. It only tells you the chance of getting six out of seven. What you need from the sign test is the chance of getting a result that extreme or more. For seven trials that means 6 or 7 as well as 0 or 1 because the test is two-tailed.

2 In a new empty cell for each possibility, repeat the previous commands but replacing the six successes with 7, 0 and 1. A good way to do this is to put the four required numbers into a column then use the cell number instead of typing in the number. E.g. if cell A1 has a 7 in it, A2 has 6, etc. then just type into cell B1 '= BINOMDIST(A1,7,0.5,0)' and then copy this down the next four cells by clicking on the small black square in the bottom right corner of the cell and dragging it down to cover the cells required.

3 Total up the four probabilities by typing '= SUM(B1 : B4)' in an empty cell. This should give you the answer 0.125. This can be interpreted as a 12.5% chance of getting a result as extreme as six out of seven or more. We have to accept the null hypothesis that there is nothing happening between the two sampling events because the value is well above $P = 0.05$. This shows the low power of the test. In fact with a sample size this small you need to get 7 out of 7 or 0 out of 7 to get a significant result (that has a probability of 0.016).

An alternative method, useful with larger samples, in Excel is to use the cumulative probability option. As this is a two-tailed test with a 0.5 chance of plus or minus, the probability of getting 1 out of 7 is the same as getting 6 out of 7. So to test the probability of getting 6 out of 7 or more extreme is equivalent to testing that of 1 out of 7. The last number of the 'BINOMDIST' function can be changed from 0 to 1 to give cumulative probability (i.e. the chance of getting that number or lower). So the probability of getting 0 *or* 1 is given by '= BINOMDIST(1,7,0.5,1)'. Doubling this value will give both the probability of getting either 6 events out of 7 or 1 out of 7.

These methods require that the number of '+' and '−' observations are known. In Excel it is easy to compare two values and tally the number of times A is larger than B using the 'IF' function. Return to the data table with paired values in columns B and C. Select cell D2. Click on 'Paste function', then 'Logical' and 'IF'. The 'Logical_test' is B2 > C2, testing whether the value in column B is greater than that in column C. The 'Value_if_true' should be 1 and 'Value_if_false' 0. Click 'OK' and paste this down the D column. A column of six '1's and one '0' should appear. The number of '1' values can be quickly totalled using 'SUM(D2 : D8)'. A word of caution: ties cause problems with this method. I suggest that they are totally ignored in the sign test. So add up the number of times A is bigger than B and then remove all instances when A = B before calculating the total number of observations.

Unpaired data

Unpaired samples or unpaired comparisons occur when a single individual is measured or tested only once. There will, therefore, be two totally separate groups of observations making up the two samples. Two groups are often obvious: e.g. males and females; kudu and eland. However, the distinction between the groups may be rather arbitrary: e.g. eastern and western; large and small. Three tests are considered below: the **independent samples *t*-test**, a **one-way analysis of variance** and the **Mann–Whitney *U* test**.

t-*test*

The independent samples *t*-test is the more usual form of the *t*-test and if the term '*t*-test' is not qualified then this is what is being referred to. The null hypothesis is that the two sets of data are the same. (Actually the null hypothesis is that the two sample means come from a population with the same true mean, μ.) The *t*-test assumes that the data are continuous, at least approximately normally distributed and that the variances of the two sets are homogeneous (i.e. the same). If possible these assumptions should be tested before the test is carried out, although this test is often incorporated into the test in statistical packages. If the two sets of observations do not have the same variance then there are ways to adjust the result of the *t*-test to compensate. It is these adjustments that are often incorporated into the test in the statistical package.

Example. The weights of five grains have been measured from each of two experimental cultivars called 'Premier' and 'Super'. Each grain has been weighed to the nearest 0.1 mg. The researcher wishes to determine whether the grain weight is the same in the two cultivars. The null hypothesis (H_0) is that the two cultivars have the same mean grain size. The alternative hypothesis (H_1) is that the two cultivars have different mean grain size.

Premier	Super
24.5	26.4
23.4	27.0
22.1	25.2
25.3	25.8
23.4	27.1

SPSS Input all of the observed data into a single column. Use another column for the labelling of the groups. This may seem wasteful but it is a much easier system when it comes to multiway analysis where each item of data will belong to several groups simultaneously.

Fig. 7.4 Arranging the data for a *t*-test in SPSS. The groups for 'cultivar' have been entered as '1' and '2' and then given labels in the 'Variable View'. The selected cell is still filled with the number '1'. Also note that the pop-up label for column two gives the full name of the variable, rather than 'grain_sz'.

The five grains from each of two cultivars of crop plant have been placed in the second column. The two cultivars have been coded as '1' or '2' in the first column. Columns have been labelled in the 'Variable View'. The package only allows restricted labels so the ideal label 'Grain size (mg)' has been shortened to 'grain_sz', although the full label can be added using the 'label' column in the 'Variable View'. The names of the groups (cultivars here) can be added using the 'values' column of the 'Variable View', and the value labels can be shown on the spreadsheet by ticking the 'Value Labels' option under the 'View' menu (Fig. 7.4).

Under the 'Analyze' menu choose 'Compare means' and then 'Independent-Samples T Test'. In the dialogue box that appears move 'grain_sz' into the 'Test Variable(s):' box by first highlighting it and then clicking on the appropriate move button. Next move 'cultivar' into the 'Grouping Variable' box. It will appear as 'cultivar(? ?)'. You need to click the 'Define Groups...' button and then input '1' in 'Group 1:' and '2' in 'Group 2:' before clicking 'Continue'. This will return you to the first dialogue box. You will see that 'cultivar(? ?)' is now 'cultivar(1 2)'. Click 'OK' to run the test.

The output will come in two parts. The first gives some information about the data like this:

Group Statistics

	CULTIVAR	N	Mean	Std. Deviation	Std. Error Mean
Grain size (mg)	Premier	5	23.740	1.218	.545
	Super	5	26.300	.806	.361

This confirms the type of test. The table then shows that the variable used was 'Grain size (mg)' and that there were two groups of 'cultivar' called 'Premier' and 'Super' each with an 'N' of 5 (number of observations). Then come some simple descriptive statistics of the two samples: their mean, 'Std. Deviation' and 'Std. Error Mean' (standard error of the mean). (Note that the full name of the data variable and the group names for '1' and '2' will only appear if they have been added in the 'Variable View'.)

The second part of the output contains the result of the t-test itself:

Independent Samples Test

		Levene's Test for Equality of Variances		t-test for Equality of Means					95% Confidence Interval of the Difference	
		F	Sig.	t	df	Sig. (2-tailed)	Mean Difference	Std. Error Difference	Lower	Upper
Grain size (mg)	Equal variances assumed	.762	.408	-3.919	8	.004	-2.560	.653	-4.066	-1.054
	Equal variances not assumed			-3.919	6.941	.006	-2.560	.653	-4.107	-1.013

First comes the 'Levene's Test for Equality of Variances'. As the t-test assumes that the two samples have equal variance SPSS sensibly tests this every time you carry out a t-test. The important figure is the last one. In this case it is 'Sig. = .408' (i.e. $P > 0.05$) so there is no evidence that the variances are unequal. If the P-value ('Sig.') is lower than 0.05 then you should be wary about using the t-test and should use the Mann–Whitney U test instead.

The output appears on two lines. The upper one, 'equal', is the standard t-test and the lower one, 'unequal', is a more conservative version that compensates for the possible problems caused by difference in variances by using a reduced value for the degrees of freedom in the test. The lower the 'Sig.' value in the Levene test, the bigger the difference between the two lines. In most cases you should not need to worry about the fact that there are two versions of the test because in most cases, as in this example, they will tell the same story.

The 't' is the actual result of the test that can be looked up on Student's t-table. 'd.f.' is the degrees of freedom of the test. This will be two fewer than the total number of observations in the two samples. The result of the Levene test determines the reduced degrees of freedom used in the lower line.

Next comes the important bit, labelled 'Sig. (2-tailed)' it is the probability that the null hypothesis is correct. This is what is usually called the P-value. In this case the *P*-value is much less than 0.05 so it is clear that the null hypothesis is extremely unlikely to be true. In fact there is only a 0.4% chance that the null hypothesis is true using the basic *t*-test, and 0.6% using the more conservative version (on the lower line).

'Mean difference' gives the value of the mean of group 1 minus the mean of group 2. It will be a negative value if group 2 is larger than group 1. 'Std. Error Difference' gives the standard error for this difference. Finally comes a '95% Confidence Interval of the Difference' column. This gives the range of difference between the two means within which 95% of samples are likely to come. In the example the 'Lower' and 'Upper' values are both negative which shows that cultivar 2, 'Super', is very likely to be heavier than cultivar 1, 'Premier'. When the *t*-test result is not significant one value will be positive and the other negative.

MINITAB There are two methods. In the first the two groups should be in two columns, in the second all the data are in one column and another column is used to identify the groups. The second method appears wasteful but it is a required strategy in more complex analysis.

Method 1. Input the data into two columns and label the columns appropriately. From the 'Stat' menu select 'Basic statistics' and then '2-sample t...'. In the dialogue box that appears select the 'Samples in different columns' option. Then place the appropriate column labels ('Premier' and 'Super' in the example) in the two boxes labelled 'First:' and 'Second:'.

Make sure the pull-down menu in the 'Options...' dialogue reads 'not equal' and the 'Confidence level:' is set at 0.95 (this means that the mean for the difference between the two means will be calculated with a 95% confidence interval). Leave the 'Assume equal variances' box unchecked. Click 'OK' to run the test.

(*Or, if the command interface is enabled, type* 'twosample c1 c2' *at the MTB > prompt in the session window.*)

The following output appears:

Two-Sample T-Test and CI: Premier, Super

Two-sample T for Premier vs Super

	N	Mean	StDev	SE Mean
Premier	5	23.74	1.22	0.54
Super	5	26.300	0.806	0.36

Difference = mu Premier - mu Super
Estimate for difference: -2.560
95% CI for difference: (-4.159, -0.961)
T-Test of difference = 0 (vs not =): T-Value = -3.92 P-Value = 0.008 DF = 6

This output confirms the test was a *t*-test and confirms the names of the two variables. It then gives summary statistics for the two groups: number of observations ('N'), mean, standard deviation and standard error of the mean. The next line gives the 95% confidence interval for the mean difference between the two groups. The references to 'mu' are to the Greek letter, μ, which is used to denote a mean. In the last line is the output for the *t*-test itself, confirming that a test of equal means is being made. The value of *t* is given as -3.92 and the important *P*-value as 0.008. This value is much less than 0.05 so we reject the null hypothesis and accept the H_1 that the two groups have different means. Inspection of the summary statistics shows that 'Super' has the larger grain size.

The degrees of freedom ('DF') is given as 6. There should be 8 degrees of freedom for the example but, as the assumption of equality of variance was not made, a correction is applied to the degrees of freedom to make the test more conservative. If the 'Assume equal variances' is checked then the example data will give 8 degrees of freedom, the same value for *t*, and a *P*-value of 0.004.

Method 2. Input all the data into a single column and use a second column to label the data. These labels should be integers. In the example it is probably best to label 'Premier' as '1' and 'Super' as '2'. Label the columns appropriately. In the example the data column should be labelled 'Grain sz' and the group codes as 'Cultivar'.

From the 'Stat' menu select 'Basic statistics' and then '2-sample t...'. In the dialogue box that appears select the 'Samples in one column' option. Then place the column with the data in the 'Samples:' box and the one with the group codes in the 'Subscripts:' box. Make sure the pull-down menu in the 'Options...' dialogue reads 'not equal' and the 'Confidence level:' is set at 0.95. Leave 'Assume equal variances' unchecked. Click 'OK' to run the test.

(*Or type* 'twot c1 c2' *(assuming data are in c1) at the MTB > prompt in the session window.*)

You get the following output:

```
Two-Sample T-Test and CI: Grain sz, Cultivar

Two-sample T for Grain sz

Cultivar    N      Mean     StDev   SE Mean
1           5     23.74      1.22      0.54
2           5     26.300    0.806      0.36

Difference = mu (1) - mu (2)
Estimate for difference:   -2.560
95% CI for difference: (-4.159, -0.961)
T-Test of difference = 0 (vs not =): T-Value = -3.92  P-Value = 0.008  DF = 6
```

Apart from the code names '1' and '2' replacing the group names the output is identical to that produced in method 1.

Excel In this case the data may be anywhere on the spreadsheet. As long as you know the cell locations of the two groups there is no problem. However, in practice, it is much easier if the data are either input exactly as in the spss example above, with one column defining the group and another containing the actual data or in two adjacent, and clearly labelled, columns.

Assuming you have input the data in an identical format to the SPSS example for the *t*-test then the data for the first cultivar is in cells b2 to b6 and for the second cultivar in cells b7 to b11. There are now two methods that may be used.

Method 1. Go to 'Paste function', select 'Statistical' and then 'TTEST'. Define the first array ('Array 1') as 'b2 : b6' and the second array as 'b7 : b11'. These arrays may be defined by selecting the box then clicking and dragging over the appropriate cells on the spreadsheet. Select 'tails' as '2' (you will nearly always require a two-tailed test) and the 'type' as '2' (this selects a standard *t*-test). Click 'OK'. The probability or *P*-value of '0.0044' will then appear in the cell.

Method 2. Type in 'TTEST' followed by, in brackets, the first and last cell of the first column separated by a colon, then the same for the second column. Then type in the number of tails in the test (usually 2) and then a 2 to ask for a standard *t*-test. In this case you would type '= TTEST(B2 : B6,B7 : B11,2,2)' and the probability will appear in the cell. Excel does not report the *t*-value.

If you are concerned that the variances of the two samples may not be equal, or you know it to be the case, then you should not use the standard *t*-test. Excel allows you to carry out a *t*-test that does not assume homogeneity of variances. This is easily accessed by using a type 3 test instead of a type 2. If you use a type 3 *t*-test with this example the probability should be reported as '0.0058'.

One-way ANOVA

Using analysis of variance (ANOVA) to determine whether just two groups have the same mean may seem like overkill. This may be the simplest use of ANOVA but it still works and gives the same answer as the *t*-test. I am of the opinion that the fact that the *t*-test is restricted to two groups makes the use of ANOVA preferable in this situation because you do not have to learn a new test when you consider more than two groups.

ANOVA has the same basic assumptions as the *t*-test: that the data are continuous, at least approximately normally distributed and that the variances of the data sets are homogeneous. These assumptions should be tested before the test is carried out. The null hypothesis is that the sets of data have the same mean. (The way that ANOVA actually approaches this is to have a null hypothesis that the variation within groups is the same as variation between groups.)

Example. In the illustrations of the use of the test for the packages I will be using the same example data set as in the *t*-test, with two samples of five observations each and each sample coming from a different cultivar of a crop plant.

The ideal ANOVA table that you would include in a write up or publication should appear as something along the lines of the table shown opposite.

SS is the sum of squares, MS is the mean square (SS/d.f.), *F* is the ratio of within group variation to between group variation (in this case it is the MS for

Source	d.f.	SS	MS	F	P
Cultivar	1	16.38	16.38	15.36	0.004**
Residual	8	8.53	1.07		
Total	9	24.92	2.77		

cultivar/residual MS). The P-value is the important one and in this case it is less than 0.05 so we reject the null hypothesis that the two cultivars have the same mean grain size. The two asterisks indicate a highly significant result and are often used when $P < 0.01$. Compare this table with the output obtained using the packages below to see where the various numbers have come from. (*Note*: the word 'residual' often appears as 'error'.)

If you wish to write this result in the text of a report the standard way would be as follows: *analysis of variance showed that the grain size of the two cultivars was significantly different ($F_{1,8} = 15.36$, $P < 0.01$).* The two subscripted numbers after the F indicate the degrees of freedom for the between group and within group variance.

SPSS As with the *t*-test all the data are placed into a single column and another column is used for the labelling of the groups (see Fig. 7.4). This may seem wasteful but it is a much easier system when it comes to more complicated analyses.

There are at least two routes to this test in SPSS. Unfortunately they give rather different outputs. I will describe them in detail below. I suggest you try both methods as the comparisons may help you understand how ANOVA tables work, and especially which parts of the tables to look at.

Method 1. Under the 'Analyze' menu choose 'Compare means' and then 'One-way ANOVA'. In the dialogue box that appears move 'grain_sz' into the 'Dependent list:' box by first highlighting it and then clicking on the appropriate move button. Next move 'cultivar' into the 'Factor' box. You can run the test now, although clicking 'Options...' allows you to request a 'Means plot' which is useful for visualization of the data, some 'Descriptive' statistics of the data and a test for 'Homogeneity of variance' (Fig. 7.5). Click 'Continue' to leave the 'Options' box, then 'OK' to run the test.

This is the minimum output you will get:

ANOVA

Grain size (mg)

	Sum of Squares	df	Mean Square	F	Sig.
Between Groups	16.384	1	16.384	15.362	.004
Within Groups	8.532	8	1.067		
Total	24.916	9			

Fig. 7.5 Using spss for one-way analysis of variance. The 'Options' dialogue box allows selection of useful additional output.

This confirms that the data is 'Grain size (mg)' (or 'grain_sz' if the variable has not been given an extra label). In the ANOVA table the top line is the important one; this is the variation between groups (i.e. between the two cultivars in this example). The second line is the variation within the groups that is being used as a comparison.

The first column is 'Sum of Squares' (often called SS), then 'd.f.' or degrees of freedom. As there are two group there is 1 degree of freedom between groups. There were five samples in each group giving two sets of 4 degrees of freedom and therefore 8 in total. Next comes a calculation column, 'Mean square' (often MS), which is the sum of squares divided by the degrees of freedom. Both SS and MS are customarily included in ANOVA tables.

Finally, comes the important bit: the F-ratio, 'F'. This is the mean square for between groups divided by that for within groups. If there is the same amount of variation between and within groups this will give an F-ratio of exactly 1. In this case the F-ratio is over 15. This value could be looked up on an F-table although spss gives you the P-value anyway, labelling it as 'Sig.'. The value is 0.004 indicating that the two groups are highly significantly different since $P < 0.01$.

One of the assumptions of ANOVA is that variances are equal. In the t-test in spss this is tested automatically. In ANOVA it is available under the 'Options...' button. Just check the box labelled 'Homogeneity-of-variance' and click 'Continue' before running the test. If you do this a little extra output appears before the ANOVA table.

Oneway

Test of Homogeneity of Variances

Grain size (mg)

Levene Statistic	df1	df2	Sig.
.762	1	8	.408

The important value is given the label 'Sig.'. The critical value is usually 0.05 and if the value given here is less than that ANOVA should not be used but a Mann–Whitney U test used instead. The value in the example is well above 0.05.

Method 2. Under the 'Analyze' menu choose 'General Linear Model' and then 'Univariate...'. In the dialogue box that appears move 'grain_sz' into the 'Dependent Variable:' box by first highlighting it and then clicking on the appropriate move button. Next move 'cultivar' into the 'Fixed Factor(s):' box. Click 'OK' to run the test.

Univariate Analysis of Variance

Between-Subjects Factors

		Value Label	N
CULTIVAR	1	Premier	5
	2	Super	5

Tests of Between-Subjects Effects

Dependent Variable: Grain size (mg)

Source	Type III Sum of Squares	df	Mean Square	F	Sig.
Corrected Model	16.384[a]	1	16.384	15.362	.004
Intercept	6260.004	1	6260.004	5869.671	.000
CULTIVAR	16.384	1	16.384	15.362	.004
Error	8.532	8	1.067		
Total	6284.920	10			
Corrected Total	24.916	9			

a. R Squared = .658 (Adjusted R Squared = .615)

The first table confirms that the data has been grouped using 'cultivar', it gives the group labels ('Value Label') and the number of observations ('N'). It then goes on to give a rather less standard ANOVA table. Remember that this output is designed to cope with many factors and therefore extra lines, which appear totally superfluous here, are useful in more complex experiments. The important line is repeated twice, although you should read along the line labelled 'CULTIVAR'. This is the variation that was labelled 'between groups' in the first method. The line labelled 'error' corresponds to the one labelled 'within groups'.

The first column is 'Type III Sum of Squares' (do not worry about type III, just report it as SS in a table), then 'd.f.' or degrees of freedom, then 'Mean Square' (the mean square value is the sum of the square value divided by the degrees of freedom). As there are two cultivars there is 1 degree of freedom. There were five samples within each cultivar giving two lots of 4 degrees of freedom and therefore 8 in total for 'error'.

Finally comes the important bit: the F-ratio, labelled 'F' here. This is the mean square for 'cultivar' divided by that for 'error'. The value of 'F' is 15.362. SPSS gives you the P-value associated with this value of 'F' and these degrees of freedom, and labels it 'Sig.'. In biology we usually look for a value of less than 0.05. Here the probability is 0.004 and indicates that the mean grain size of the two cultivars are significantly different. The output of this method is certainly not appropriate for inclusion in a write up as it stands; create a table in the form given earlier in the example.

There are even more options to try in method 2. The 'Homogeneity tests' option is one that should usually be selected as it tests whether the data are appropriate for ANOVA. An output will appear similar to that in the data output at the top of p. 97. If 'Compare main effects' is selected with 'cultivar' in the 'Display Means for:' box, the following table will appear:

Pairwise Comparisons

Dependent Variable: Grain size (mg)

(I) CULTIVAR	(J) CULTIVAR	Mean Difference (I-J)	Std. Error	Sig.[a]	95% Confidence Interval for Difference[a]	
					Lower Bound	Upper Bound
Premier	Super	-2.560*	.653	.004	-4.066	-1.054
Super	Premier	2.560*	.653	.004	1.054	4.066

Based on estimated marginal means

*. The mean difference is significant at the .05 level.

a. Adjustment for multiple comparisons: Least Significant Difference (equivalent to no adjustments).

This is similar to the mean difference information given at the end of the t-test. It shows the differences between the mean values for the two factor levels (groups, or 'cultivars' in the example), gives a standard error for the difference, a significance for the difference ('Sig.') and then 95% confidence intervals for the difference. If both 'Lower Bound' and 'Upper Bound' values have the same sign then the two factor levels are significantly different from each other.

MINITAB As with the t-test, there are two ways of inputting the data. In the first the two groups should be in two columns, in the second all the data are in one column and another column is used to identify the groups. The second method appears wasteful but arranging the data in this way is required in more complex analyses.

Method 1. Input the data from the two groups into two separate columns and label appropriately. From the 'Stat' menu select 'ANOVA' then 'One-way (Unstacked)'. In the dialogue box ensure that both columns are in the 'Responses (in separate columns):' box. Click 'OK'.

(*Or, if the command interface is enabled, type* 'aovo c1 c2' *at the MTB > prompt in the session window.*)

You get this output:

```
One-way ANOVA: Premier, Super

Analysis of Variance
Source      DF        SS        MS        F         P
Factor       1     16.38     16.38     15.36     0.004
Error        8      8.53      1.07
Total        9     24.92
                                  Individual 95% CIs For Mean
                                  Based on Pooled StDev
Level        N      Mean     StDev   ---------+---------+---------+-------
Premier      5    23.740     1.218   (------*------)
Super        5    26.300     0.806                   (------*------)
                                  ---------+---------+---------+-------
Pooled StDev =      1.033             24.0      25.5      27.0
```

The first part of the output gives an ANOVA table in a fairly standard form (see the 'ideal' version in the example section). The highly significant *P*-value of 0.004 means we reject the null hypothesis that the two cultivars have the same mean. However it does not help us decide which group has the higher mean. The second section provides information about the two groups: number of observations ('N'), mean and standard deviation and then a rather primitive, but useful, graphical representation of the two group means and 95% confidence intervals of the means. In this example the confidence intervals of the two groups do not overlap (confirming that they are significantly different from each other) and clearly 'Super' has the greater mean.

Method 2. Input all the observations into a single column. Use a separate column for coded labels for the two groups. In this example you would call the first column 'Grain sz' and the second 'Cultivar'. You must use integers as codes for groups. From the 'Stat' menu select 'ANOVA' then 'One-way...'. Move the observed data into the 'Response:' box and the group codes into the 'Factor:' box. Do not worry about the comparisons button for only two groups, it will become useful when you have more than two groups. Click 'OK'.

(*Or type* 'onew c1 c2' *at the MTB > prompt in the session window.*)

The output is identical to that from method 1 except that the two groups will be labelled with numbers rather than names.

Excel There is no direct method unless you have installed the 'Data analysis toolpak' add-in.

1 Put the data into an Excel spreadsheet. This can be done in several ways: either exactly as in the example with one column for each cultivar and the

Fig. 7.6 One-way (single-factor) analysis of variance using Excel. The area of the spreadsheet has been selected and is highlighted with a dashed line. Currently the output options will put the output onto a fresh sheet in the same workbook.

names in the first row, or with all the data in a single column and an extra column having the names of the cultivars. (I will assume here that the data are in two columns A and B with labels in cells A1 and B1).

2 Select the 'Tools' menu and select 'Data analysis…' (if this option does not appear you need to run Excel set-up and add the 'Data analysis toolpak' add-in). Select 'Anova: Single Factor' and click 'OK'.

3 A dialogue box will appear. If the cursor is flashing in the 'Input range' box you can select the cells you wish to use for the analysis by clicking and dragging in the main sheet. If you select from cell A1 to the end of the data 'A1 : B6' should appear in the box. Alternatively you can just type in 'A1 : B6'. The option 'columns' should be selected because the data are indeed in two columns representing different groups. The tick box 'Labels in the first row' should be checked as the cultivar names are in the first row. Leave the 'Alpha' at 0.05 as this is the significance level which is chosen and $P < 0.05$ is the usual level (Fig. 7.6).

4 At the bottom of the dialogue box is a section allowing you to determine where the output will appear. The default option is 'New Worksheet Ply' which means that the output will appear in a different sheet. If you want the output to appear on the same sheet as the data then you need to put a cell number in the 'Output range' box that will determine where the top left cell of the output will start.

5 Click on 'OK' and the following output will appear:

Anova: Single Factor				
SUMMARY				
Groups	*Count*	*Sum*	*Average*	*Variance*
Premier	5	118.7	23.74	1.483
Super	5	131.5	26.3	0.65
ANOVA						
Source of Variation	*SS*	*df*	*MS*	*F*	*P-value*	*F crit*
Between Groups	16.384	1	16.384	15.3624	0.004422	5.317645
Within Groups	8.532	8	1.0665			
Total	24.916	9				

The first section gives summary information about the two groups of observations confirming the number of observations, sum, mean and variance.

The second section presents a rather standard ANOVA table: 'SS' is the sum of squares; 'd.f.' degrees of freedom (there were 10 observations in all giving a total of 9 degrees of freedom and there were two groups giving 1 degree of freedom for the 'between groups' variation); 'MS' is the mean square (= SS/d.f.); 'F' is the ratio of the variation between groups/within groups; and 'P-value' is the important value as it shows the probability that the two cultivars have the same mean grain size. In this case $P = 0.0044$ so we reject the null hypothesis that the two cultivars have the same mean grain size. The final value 'F crit' is not usually quoted on an ANOVA table. It is the value of F required to achieve a $P = 0.05$ with the degrees of freedom in this particular test. If alpha had been set at 0.01 in the ANOVA dialogue box then the 'F crit' value would be 11.25863.

Mann–Whitney U *test*

This test, also widely known as the Wilcoxon–Mann–Whitney test and less widely as the Wilcoxon rank sum W test, is the non-parametric equivalent of the independent samples t-test. It can only be used to test two groups. However, unlike the t-test and one-way ANOVA it does not make assumptions about homogeneity of variances or normal distributions. It is a typical 'rank' test, meaning that the raw data are converted into ranks before the test is carried out. The advantage of this is that it is ideal for situations where the highest value went off the scale or if extreme values are making the t-test undesirable.

The Mann–Whitney U test is less powerful than a t-test or one-way ANOVA, but you are less likely to find a significant result when there is no real difference. However, the lack of assumptions it makes about the distribution of the data make it the preferred test in most cases.

Example. In the package illustrations for this test I use the same data set as for the t-test and one-way ANOVA.

SPSS As with the *t*-test all the data are placed into a single column and another column is used for the labelling of the groups using integer code numbers that can be labelled as appropriate in the 'Variable View', 'Values' column. Under the 'Analyze' menu, choose 'Nonparametric Tests' and then select '2-Independent Samples...'. This will bring up a dialogue box for four different non-parametric tests. By default the 'Mann–Whitney *U* test' box should be selected. Put 'grain_sz' into the 'Test Variable List' box by selecting it and then moving it across. Then select 'cultivar' as the 'Grouping Variable'. It will appear as 'culti-var(? ?)'. You need to click the 'Define Groups...' button and then input '1' in 'Group 1' and '2' in 'Group 2' before clicking 'Continue'. This will return you to the first dialogue box. You will see that 'cultivar(? ?)' is now 'cultivar(1 2)'. Click 'OK' to run the test.

The output should appear as follows:

Mann-Whitney Test

Ranks

	CULTIVAR	N	Mean Rank	Sum of Ranks
Grain size (mg)	Premier	5	3.20	16.00
	Super	5	7.80	39.00
	Total	10		

Test Statistics[b]

	Grain size (mg)
Mann-Whitney U	1.000
Wilcoxon W	16.000
Z	-2.410
Asymp. Sig. (2-tailed)	.016
Exact Sig. [2*(1-tailed Sig.)]	.016[a]

a. Not corrected for ties.

b. Grouping Variable: CULTIVAR

This confirms the test being carried out. It also confirms that the variable 'Grain size (mg)' was divided into two groups by the variable 'cultivar' labelled as 'Premier' and 'Super'. Next comes some summary information. The two groups of 'cultivar' each have five cases ('N') giving 10 in total. The mean rank position of the two groups, with the smallest value given rank 1 and the largest rank 10, is also given. The 'Sum of Ranks' is given as it is an important part of the Mann–Whitney calculation, although it is not useful here. Even at this point it is clear that there is a difference between the two cultivars as the mean ranks are different.

The output from the test itself comes in a separate table. The first two rows are the output values from the Mann–Whitney and Wilcoxon versions of the test (U and W, respectively). 'Z' is the statistic usually looked up in a table. Finally, the rows labelled 'Asymp. Sig. (2-tailed)' and 'Exact Sig. [2*(1-tailed Sig.)]' give the crucial P-value. It is this value that you are interested in. Is it less than 0.05? In this example it is and we reject the null hypothesis that the two groups have the same median.

MINITAB Unlike the t-test there is only one way of inputting the data: put the data into two columns, one for each group, and label accordingly. From the 'Stat' menu select 'Nonparametrics' then 'Mann Whitney...'. In the dialogue box move the two columns into the 'First sample:' and 'Second sample:' boxes (it does not matter which way round they are). Leave the other settings at their defaults of '95.0' and 'not equal'. Click 'OK'.

(Or, if the command interface is enabled, type 'Mann c1 c2' at the MTB > prompt in the session window.)

You get the following output:

```
Mann-Whitney Test and CI: Premier, Super

Premier    N =   5      Median =      23.400
Super      N =   5      Median =      26.400
Point estimate for ETA1-ETA2 is       -2.500
96.3 Percent CI for ETA1-ETA2 is (-4.301,-0.701)
W = 16.0
Test of ETA1 = ETA2   vs   ETA1 not = ETA2 is significant at 0.0216
The test is significant at 0.0212 (adjusted for ties)
```

The output confirms the test carried out. Then it gives some summary information about the two groups: the number of observations ('N') and the median value. The next two lines give information about the difference between the two groups and quote a range for the difference. Then comes a value for the test statistic 'W', given as 16.0 in the example. The final two lines are the most important. They give the probability of the two medians being equal. The first probability ($P = 0.0216$ in the example) does not account for tied data (two or more observations with exactly the same value) while the bottom line does. If there are several ties in the data then the second value will be higher and the second P-value for the test becomes more conservative.

In the example both versions give a $P < 0.05$ so we reject the null hypothesis that the two groups have the same median value (i.e. in the example, the grain size for the two cultivars is different).

If the two significance results are either side of 0.05 then you will have to consider that a marginal result for the test suggests that you need to collect more data or measure with more precision to avoid tied data.

Excel There is no direct method even with the 'Data analysis toolpak' installed.

Do the observations from more than two groups differ?

The groups can be either repeated (related) samples or they can be independent. Repeated measures are considered first.

Repeated measures (a.k.a. related samples, matched samples)

This is an extension from paired data and occurs when a single individual or site is tested three or more times. A common example is in a 'before, during and after' design. Another possible use occurs when an individual is, or individuals of a clone are, divided and then subjected to three or more treatments. Two tests are considered below: the **Friedman test** and **repeated measures** ANOVA.

Friedman test (for repeated measures)

The Friedman test is a non-parametric analogue of a two-way ANOVA. It makes no assumptions about the distribution of the data, only that it is measured on an ordinal scale. The Friedman test is appropriate only if there is a single observation taken for each combination of factor levels. For repeated measures one of the factors must represent the repeat level, perhaps minutes, days or a measure of before, during and after a procedure. Then the second grouping variable will be a standard factor such as region, species or treatment type. The null hypothesis is that observations in the same group (factor level) have the same median values. If the null hypothesis is rejected it shows that at least two groups have different medians, although it does not show which groups they are. Inspection will reveal which are the likely candidate groups (i.e. those with the highest and lowest medians).

It is the case for both SPSS and MINITAB that the test has to be carried out twice if you wish to test both the conventional factor and whether the different sampling events have different median values.

The Friedman test is much less powerful than an equivalent ANOVA test if the data are normally distributed but makes fewer assumptions about the data so it is 'safer'.

Example. The Friedman test could be used if the data comprises the number of cyanobacterial cells in 1 mm^3 of water from 6 ponds, with samples taken on four different days and only one sample taken each day from each pond.

			Pond			
Day	A	B	C	D	E	F
1	130	125	350	375	225	235
2	115	120	375	200	250	200
3	145	170	235	275	225	155
4	200	230	140	325	275	215

Note: there is only one observation for each pond/day combination.

SPSS Input the data using one column for each factor level (pond) in this example. Label the columns appropriately. Make sure that each row corresponds to a level (group) of the repeated measure (e.g. if different times make sure that all the 'before' measure are in the first row, 'during' in the second and so on).

From the 'Analyze' menu select 'Nonparametric Tests' then 'K-related samples...'. In the dialogue box move all the columns containing data into the 'Test variables:' box. Make sure that the 'Friedman' box is checked (it is usually selected by default). If you want any summary statistics about each of the groups you should click the 'Statistics...' button and check 'Descriptives' and click 'Continue' before you click 'OK'.

You should get the following output:

Friedman Test

Ranks

	Mean Rank
A	1.50
B	2.50
C	4.25
D	5.38
E	4.25
F	3.13

Test Statistics[a]

N	4
Chi-Square	11.259
df	5
Asymp. Sig.	.046

a. Friedman Test

The output confirms the test. Then it gives some rather unhelpful information about the mean rank of the observations in the different samples. The test output appears in the second table. First, the number of repeated or related samples is confirmed (four in the example). Then a test statistic is given followed by the degrees of freedom (one less than the number of groups being tested, so 5 d.f. from six ponds in the example). Finally the *P*-value is given, labelled as 'Asymp. Sig.'. If this value is less than 0.05 you reject the null hypothesis that the groups have the same median. In the example the *P*-value is 0.046. This value is just less than the critical 0.05 level. So we reject the null hypothesis that the ponds have the same median concentration of cyanobacteria.

To test another possible null hypothesis that the days all have the same median cell concentration would make the test a two-way analysis (see p. 126).

MINITAB Put the data into a single column. Then use the next two columns for integer codes of the two grouping variables. These codes must be integers but need not be consecutive.

From the 'Stat' menu select 'Nonparametric' then 'Friedman...'. In the dialogue box put the data in the 'Response' block, the conventional factor into the 'Treatment:' box and the repeated measure (perhaps a time, as in the example) into the 'Block:' box. Click 'OK'.

(*Or, if the command interface is enabled, and assuming the observations are in column 1, the main factor in column 2 and the repeat factor in column 3, type* 'Friedman c1 c2 c3' *at the MTB > prompt in the session window.*)

You get the following output:

Friedman Test: cells versus pond, day

```
Friedman test for cells by pond blocked by day

S = 11.18   DF = 5   P = 0.048
S = 11.26   DF = 5   P = 0.046 (adjusted for ties)

                      Est      Sum of
pond           N     Median    Ranks
1              4     150.52      6.0
2              4     166.35     10.0
3              4     295.10     17.0
4              4     293.85     21.5
5              4     240.10     17.0
6              4     192.19     12.5

Grand median   =     223.02
```

Note: I coded the six ponds in the example with the numbers 1 to 6.

The output first confirms the test carried out. The first two rows give the results of the test, first with no correction for tied observations. The sample statistic for the test is 'S' and the 'DF' is the degrees of freedom (one less than the number of factor levels); there were six ponds in the data set so 5 degrees of freedom. Then there is a *P*-value associated with the 'S' result. In the example the *P*-value is just less than 0.05 so we reject the null hypothesis that the six ponds have the same median number of cyanobacterial cells.

After the test results come some summary statistics about the groups giving the number of observations ('N'), the median and the total rank position of the observations in the whole data set ('Sum of Ranks'). The 'Grand median' is the median value from the whole data set. By inspection you can determine that at least ponds 1 (A) and 4 (D) have significantly different medians as they are the extreme groups.

If you wish to determine whether the different sampling events have an effect on the median observations then you should repeat the test reversing the factors in the 'Treatment:' and 'Block' boxes. This makes the test into a 'two-way' analysis (see p. 122).

(*Or type* 'Friedman c1 c3 c2' *at the MTB > prompt in the session window.*)

Excel There is no direct method even with the 'Data analysis toolpak' installed.

Repeated measures ANOVA

A two-way analysis of variance may also be applied to this sampling design of only one measure for each combination of factor levels. The problem is that ANOVA makes the assumption that each of the factor levels is independent of all others. In a repeated measures design, where a measurement is taken from an experiment on days 1, 2 and 3, the 3 days cannot be used as factor levels because measurements taken on day 2 are not independent of those already taken on day 1. The factor levels are also not equally dependent as day 1 is likely to be more similar to day 2 that it will be to day 3. There are ways around this problem and they usually entail a serious reduction in the degrees of freedom applied to the analysis.

ANOVA is a parametric test and therefore makes assumptions about the data. The data should be: continuous, normally distributed and with equal variances for each factor. In practice it is often impossible to test whether this is true because there are so few observations. Therefore it is usual to make a value judgement about whether the data are likely to conform to the assumptions.

In the example used above in the Friedman test the data of cells/mm^3 is measured to the nearest 5 and is therefore discontinuous. However, there are more than 30 possible values so it would usually be acceptable to use ANOVA.

This test is only a special case of a two-way ANOVA where one of the factors defines the level of repeated sampling and the other the individuals or sampling sites. You could treat it as a two-way ANOVA (usually without replication) — see p. 126 and below.

SPSS This test is reached from 'Analyze', then 'General Linear Model' and then 'Repeated measures'. The package will not carry out the task using 'Repeated measures' if, as in the example, there is only one observation in each measure and only one individual ('pond' in the example) for each factor level.

However, if there are two or more individuals in each factor level, say if the ponds are grouped into 'ephemeral' and 'permanent', then the test can be carried out. Arrange the data so each measurement event is in a column (e.g. 'day') and each individual is on a separate row. There should be an additional column for the main effect (e.g. 'pondtype'). Use 'Analyze', then 'General Linear Model' and then 'Repeated measures'. In the dialogue box insert the name of the repeated visit (e.g. 'day') and the number of times measured (four in the example); then click 'Define'.

In the next dialogue, move the measurement columns into the 'Within-Subjects Variables' box. There should be as many columns selected as you chose levels in the first window. Move the main effect to 'Between-Subjects Factor(s)'. Click 'OK'.

Masses of output is produced, reflecting the debates in the statistical literature about the most appropriate way to deal with repeated measures designs. At the bottom is an ANOVA-style table. The row labelled with the name of the main effect in the first column ('Source') is the important one. Read the 'Sig.' value in the final column.

MINITAB There is no repeated measure option in MINITAB, although a two-way ANOVA (see p. 126) can be carried out. In a repeated measures design the F-ratio should be checked for significance with fewer degrees of freedom than for a conventional two-way ANOVA. This can be done by looking in statistical tables, or by using Excel as detailed below.

Excel A two-way ANOVA (see p. 131) can be carried out if the design is fully balanced with no missing data, although the F-ratio should be checked for significance with fewer degrees of freedom than for a conventional two-way ANOVA. This can be done with the function 'FDIST'. Use the 'Paste function' button, then 'Statistical', or type directly 'FDIST(x,y,z)' where 'x' is the F-ratio reported in the ANOVA output, 'y' is the number of degrees of freedom for the main effect (5 d.f. from six ponds in the example) and 'z' is the number of degrees of freedom in the error. It is the value of 'y' (degrees of freedom of the main effect) that should be reduced.

Independent samples

This is the more usual type of analysis and occurs when a single individual or site is measured or tested only once. There will, therefore, be three or more totally separate groups of observations. Groups are often obvious: e.g. pig, sheep, cow and horse. However, the distinction between the groups may be rather arbitrary: e.g. dividing samples by altitude bands or ranges of linear distance from a release point. Two tests are considered below: **one-way analysis of variance** and the **Kruskal–Wallis test**. If you get a significant result from one of these tests that is not the end of the story as it will not tell you which groups are different from which. A *post hoc* test of some kind (meaning 'after this') is required to allow you to interpret the results. Some of the many *post hoc* **tests** available are also described.

One-way ANOVA

The *t*-test is restricted to use with only two samples. When there are more than two groups the only sensible option for a parametric test is to use analysis of variance (ANOVA). This is still a very simple use of ANOVA but it works extremely well and is a test that every biologist should feel comfortable with.

ANOVA for three or more groups makes the same assumptions as for two groups: that the data are continuous, at least approximately normally distributed and that the variances of each group/sample are homogeneous. These assumptions should be tested before the test is carried out. The null hypothesis being tested is that each set of data has the same mean. (The way that ANOVA approaches this is to have a null hypothesis that the variation within groups is the same as the variation between groups.)

Example. A researcher has grain size data from three cultivars and wishes to determine whether there are any differences between them. The null hypothesis is that all three have the same mean (or that all three samples are taken from populations with the same mean). If there is a significant result (i.e. if $P < 0.05$) this indicates that at least one pair has different means, it does not tell you which pair. For convenience, in this example I will use the same example data set as in the *t*-test but with an additional third sample of five observations from a cultivar called 'Dupa' which is coded as number 3. I would normally code the groups in alphabetical order. In this case as a third set of data has been added to a previously used set I have retained the labels '1' and '2' for those groups.

Cultivar name	Cultivar code	Grain size (mg)
Premier	1	24.5
	1	23.4
	1	22.1
	1	25.3
	1	23.4
Super	2	26.4
	2	27.0
	2	25.2
	2	25.8
	2	27.1
Dupa	3	25.5
	3	25.7
	3	26.8
	3	27.3
	3	26.0

If the data are analysed using ANOVA on a statistics package the output will not be directly suitable for presentation in a report. The ANOVA table that you would include in a write up or publication should appear as something along the lines of the table below:

Source	d.f.	SS	MS	F	P
Cultivar	2	21.5	10.8	11.9	0.0014
Residual (error)	12	10.9	0.91		
Total	14	32.4			

Here, d.f. is degrees of freedom, SS the sum of squares, MS the mean square (sum of squares divided by degrees of freedom) and *F* the *F*-ratio (mean square of the main effect divided by the mean square of the error). The best way to test that you are reading the output from your package correctly is to compare this table with the output obtained using the package.

SPSS As with many statistical tests all the data are placed into a single column and another column is used for the labelling of the groups. This may appear to double the effort required but it is a much easier system when it comes to multi-way analysis. *Post hoc* tests are important when trying to interpret the output from ANOVA. They are considered later.

As with the two-sample tests there is more than one route to this test in SPSS that give the same results but with rather different outputs. I will consider two below in detail.

Method 1. Under the 'Analyze' menu choose 'Compare means' and then 'One-way ANOVA...'. In the dialogue box that appears move 'grain_sz' into the 'Dependent list:' box by first highlighting it and then clicking on the appropriate move button. Next move 'cultivar' into the 'Factor' box. (Ignore the 'Post Hoc' button for the moment, although you should always select a *post hoc* test when there are more than two groups. I will consider *post hoc* testing on p. 116.) Click 'OK' to run the test.

The following will appear in the output window:

Oneway

ANOVA

GRAIN_SZ

	Sum of Squares	df	Mean Square	F	Sig.
Between Groups	21.509	2	10.755	11.879	.001
Within Groups	10.864	12	.905		
Total	32.373	14			

This confirms the test as 'Oneway, ANOVA', tells you that the observed data is 'grain_sz' and then goes on to a standard ANOVA table. As is always the case in one-way ANOVA, the top line of the table is the important one: this is the variation between groups (i.e. between the three cultivars in this example). The second line is the variation within the groups that is being used as a comparison.

The first column is the 'Sum of Squares' (often just 'SS'), then comes 'd.f.' or degrees of freedom. As there are three groups ('cultivars') there are 2 degrees of freedom between groups (i.e. $3 - 1 = 2$). There were five samples within each group giving three lots of 4 degrees of freedom and therefore 12 in total. Next comes the calculation column 'Mean Square' (usually just 'MS'). The mean square value is the sum of square value divided by the degrees of freedom.

Finally comes the important bit: the 'F' (F-ratio). This is the mean square for between groups divided by that for within groups. If there is the same amount of variation between and within groups this will give an F-ratio of exactly 1. In this case the F-ratio is 11.879. This value could be looked up on an F-table although SPSS gives you the P-value anyway, labelling it as 'Sig.'. The value is 0.001 indicating that *at least* two of the groups have means that are highly significantly different. There are three possible pairs with three groups: 1 and 2; 1 and 3; 2 and 3. It is the *post hoc* test that tells you which pairs are different from which **not the** ANOVA *test itself.*

More complicated ANOVA tests are not possible under the 'Compare Means' menu so it may be better to become familiar with method 2.

One of the assumptions of ANOVA is that variances are equal. In the t-test in SPSS this is tested automatically. In ANOVA it is available under the 'Options...' button. Just check the box labelled 'Homogeneity of Variance' and click 'Continue' before running the test. If you do this a little extra output appears after the ANOVA table.

Test of Homogeneity of Variances

GRAIN_SZ

Levene Statistic	df1	df2	Sig.
.679	2	12	.526

The important value is given the label 'Sig.'. The critical value is usually 0.05 and if the value given here is less than that ANOVA should not be used and a Kruskal–Wallis test should be considered instead (see p. 119).

Method 2. Under the 'Analyze' menu choose 'General Linear Model' and then 'Univariate...'. In the dialogue box that appears move the data column ('grain_sz' in the example) into the 'Dependent Variable:' box by first highlighting it and then clicking on the appropriate move button. Next move the grouping variable ('cultivar' in the example) into the 'Fixed Factor(s)' box. Either click 'OK' to run the test now or click on 'Options...' and select 'Homogeneity tests' first.

Univariate Analysis of Variance

Between-Subjects Factors

		N
CULTIVAR	1	5
	2	5
	3	5

Levene's Test of Equality of Error Variances[a]

Dependent Variable: GRAIN_SZ

F	df1	df2	Sig.
.679	2	12	.526

Tests the null hypothesis that the error variance of
the dependent variable is equal across groups.

a. Design: Intercept+CULTIVAR

Tests of Between-Subjects Effects

Dependent Variable: GRAIN_SZ

Source	Type III Sum of Squares	df	Mean Square	F	Sig.
Corrected Model	21.509[a]	2	10.755	11.879	.001
Intercept	9702.817	1	9702.817	10717.397	.000
CULTIVAR	21.509	2	10.755	11.879	.001
Error	10.864	12	.905		
Total	9735.190	15			
Corrected Total	32.373	14			

a. R Squared = .664 (Adjusted R Squared = .608)

This output confirms the test used. Then comes a table containing only the number of observations for each factor level. The second table is the output requested in 'Options...' for 'Homogeneity tests'. A Levene test has the null hypothesis that each factor level has the same variance. In the example the 'Sig.' for this test is well above 0.05, so there is no reason to reject this null hypothesis. If the 'Sig.' is below 0.05 then an alternative test, such as the Kruskal–Wallis test, should be considered.

Finally comes a rather less standard ANOVA table labelled as 'Tests of Between-Subjects Effects'. This output is designed to cope with many factors and therefore extra lines, which appear superfluous here, are useful in more complex experiments. The important line is repeated twice, although you should read along the line labelled 'CULTIVAR' (or whatever your main effect is labelled as). This is the test of the null hypothesis that between-group (factor levels) variation is the same as within-group variation. The line labelled 'Error' is the residual variation or within-group variation.

The second column is 'Type III Sum of Squares' (do not worry about type III, just report it as SS in a table), then 'd.f.' or degrees of freedom, then 'Mean Square' (the mean square value is the sum of square value divided by the degrees of freedom). As there are three cultivars there are 2 degrees of freedom. There were five samples within each cultivar giving three sets of 4 degrees of freedom and therefore 12 in total for 'Error'.

Finally comes the important bit: the *F*-ratio, labelled 'F' here. This is the mean square for 'cultivar' divided by that for 'error'. The value of 'F' is 11.879. SPSS gives you the *P*-value associated with this value of 'F' and these degrees of freedom and labels it 'Sig.'. We look for a value less than 0.05. Here the probability is 0.001 and indicates that the mean grain sizes of the two cultivars are significantly different. (Note that SPSS does have a habit of reporting significance of 0.000. This should be reported as $P < 0.001$.)

The output table should be revised along the lines of the table given at the top of this section before it is included in a write up or report.

If you select the main effect and add it to 'Display Means for;' and check 'Compare main effect' you get an output as follows:

Estimates

Dependent Variable: GRAIN_SZ

CULTIVAR	Mean	Std. Error	95% Confidence Interval	
			Lower Bound	Upper Bound
1	23.740	.426	22.813	24.667
2	26.300	.426	25.373	27.227
3	26.260	.426	25.333	27.187

Pairwise Comparisons

Dependent Variable: GRAIN_SZ

(I) CULTIVAR	(J) CULTIVAR	Mean Difference (I-J)	Std. Error	Sig.[a]	95% Confidence Interval for Difference[a]	
					Lower Bound	Upper Bound
1	2	-2.560*	.602	.003	-4.227	-.893
	3	-2.520*	.602	.004	-4.187	-.853
2	1	2.560*	.602	.003	.893	4.227
	3	4.000E-02	.602	1.000	-1.627	1.707
3	1	2.520*	.602	.004	.853	4.187
	2	-4.000E-02	.602	1.000	-1.707	1.627

Based on estimated marginal means

*. The mean difference is significant at the .05 level.

a. Adjustment for multiple comparisons: Sidak.

First a table gives the mean values for each level of the main effect (here still coded only as '1', '2' and '3'). It also gives 95% confidence interval ranges. The second table gives pairwise comparisons for each pair of factor levels. The differences in the means and the 95% confidence intervals for the differences are also given. In this example I have selected the rather conservative Sidák method for pairwise comparisons (discussed below in the section on *post hoc* testing). In this example the differences between groups 1 and 2 and between 1 and 3 are significant ($P = 0.003$ and 0.004, respectively), but the difference between groups 2 and 3 is not.

MINITAB There are two routes to this test in MINITAB. You can either input all the data into a single column with a second column to code the groups (stacked) or you can put each group into a separate column (unstacked).

Method 1. Input all the data into a single column and use a second column to label the cultivars with integers (as in the example). Label the columns appropriately. From the 'Stat' menu select 'ANOVA' then 'Oneway...'. Move the observed data into the 'Response:' box and the group codes into the 'Factor:' box. The 'Comparisons...' button gives you access to the *post hoc* tests that are considered separately. Click 'OK' to run the test.

(Or, *if the command interface is enabled, type* 'Oneway c1 c2' *(assuming the data are in c1 and the group codes in c2) at the MTB > prompt in the session window.*)

You get the following output:

```
One-way ANOVA: Grain sz versus Cultivar

Analysis of Variance for Grain sz
Source      DF        SS        MS        F        P
Cultivar     2    21.509    10.755    11.88    0.001
Error       12    10.864     0.905
Total       14    32.373
                                    Individual 95% CIs For Mean
                                    Based on Pooled StDev
Level        N      Mean     StDev   ----------+---------+---------+------
1            5    23.740     1.218   (-------*-------)
2            5    26.300     0.806                     (-------*-------)
3            5    26.260     0.764                    (-------*-------)
                                    ----------+---------+---------+------
Pooled StDev =     0.951              24.0      25.2      26.4
```

The first part of the output confirms the test carried out and gives an ANOVA table in a standard form (see the 'ideal' version in the example section). The highly significant *P*-value (much less than 0.05) means we reject the null hypothesis that the cultivars have the same mean. However it does not help us decide which groups are different from which. The second section provides information about the three groups: number of observations ('N'), mean and standard deviation, and then a rather primitive graphical representation of the three group means and 95% confidence intervals of the means. In this example the confidence intervals of group ('Level') 1 does not overlap with groups 2 and 3 (confirming that they are significantly different from each other). The *post hoc* test will confirm this conclusion derived from inspection of the data (see p. 116).

Method 2. There is no option for *post hoc* testing with this method, making method 1 far superior. Input the data with one column for each of the groups (cultivars in the example) and label the columns appropriately. From the 'Stat' menu select 'ANOVA' then 'Oneway (unstacked)...'. In the dialogue box move all groups into the 'Responses (in separate columns):' box. Then click 'OK'. The output is identical to method 1 but with the column names appearing instead of the integer codes for the factor levels.

Excel There is no direct method unless you have installed the 'Data analysis toolpak' add-in. The method is essentially the same as for two groups.

1 Input the data onto an Excel spreadsheet. This can be done in several ways: either with one column for each cultivar and the names in the first row, or, as in the example, with the data from one cultivar in one row with an extra column having the names of the cultivars. (I will assume here that the data are in three columns A, B and C with the names of the cultivars in cells A1, B1 and C1).

2 Select the 'Tools' menu and select 'Data analysis...' (if this option does not appear you need to run Excel set-up and add the 'Data analysis toolpak' add-in). Select 'Anova: Single Factor'.

3 A dialogue box will appear. If the cursor is flashing in the 'Input range' box you can select the cells you wish to use for the analysis by clicking and dragging in the main sheet. If you select from cell A1 to the end of the data 'A1 : C6' should appear in the box. Alternatively you can just type in 'A1 : C6'. The option 'Columns' should be selected because the data are indeed in two columns representing different groups. The tick box 'Labels in First Row' should be checked as the cultivar names are in the first row. Leave the 'alpha' at 0.05 as this is the significance level which is chosen and $P < 0.05$ is the usual level used in biology.

4 At the bottom of the dialogue box is a section allowing you to determine where the output will appear. The default option is 'New Worksheet Ply' which means that the output will appear in a different sheet. If you want the output to appear on the same sheet as the data then you need to put a cell number in the 'Output range' box that will determine where the top left cell of the output will start.

5 Click on 'OK' and the following output will appear:

Anova: Single Factor

SUMMARY

Groups	Count	Sum	Average	Variance
Premier	5	118.7	23.74	1.483
Super	5	131.5	26.3	0.65
Dupa	5	131.3	26.26	0.583

ANOVA

Source of Variation	SS	df	MS	F	P-value	F crit
Between Groups	21.50933	2	10.75467	11.87923	0.001428	3.88529
Within Groups	10.864	12	0.905333			
Total	32.37333	14				

The important value is the 'P-value'. If this is less than 0.05 we reject the null hypothesis that all three cultivars have grains with the same mean size. In this case the value is well below that level at $P = 0.0014$ so we can say that the result is highly significant. However, this just gives the probability that two of the cultivars have different means — it does not show which ones. Inspection of the 'Summary' table shows that 'Super' and 'Dupa' have very similar means but 'Premier' is different from them. There is no *post hoc* test available in Excel.

See the section on the two groups above for a description of the rest of the output (p. 101).

Post hoc *testing: after one-way* ANOVA

One of the commonest errors of omission I see is a one-way ANOVA carried out on, say, four groups with a *P*-value well under 0.05 and then no further invest-igation of which groups are different from which. It is important to realize that a significant result in the ANOVA will only show that at least one pair of the groups is significantly different. It does not identify which pair(s). When there are three groups there are only three possible pairs, with four groups this rises to six pairs and with five groups there are ten. *Post hoc* tests help you to make sense of this large number of possible comparisons by actually identifying which groups are significantly different from which.

The only problem is that the number of methods at your disposal for this task is overwhelming. For instance in SPSS you are offered a choice of seven different methods for answering the same question and these are only a subset of the number available.

The two tests I suggest you look for, and they will almost invariably give you the same results, are the **least significant difference (LSD) test** (a.k.a. Fisher LSD test) and the **Student–Newman–Keuls (SNK) test**. The LSD test should *only* be carried out when the ANOVA result is significant. It uses the logic that if only significant results are examined there is no need to reduce the critical *P*-value (alpha) below 0.05 for the pairwise comparisons.

More conservative *post hoc* tests include the **Bonferroni method**, which is a general method for reducing the critical level required for significance below 0.05, achieved by dividing this value by the number of comparisons made, and the **Dunn–Sidák method**, which is another general method of setting the critical value needed for significance (usually set to 0.05) when many pairwise compar-isons are carried out at the same time (see p. 193).

It is conventional to show the results of a significant ANOVA by arranging the groups into ascending order of mean and then drawing lines under groups that are *not* significantly different. Therefore for the example I have been using here:

```
Cultivar 1    Cultivar 3    Cultivar 2
              _____
```

Another very commonly employed technique for showing the results of a *post hoc* test is to use a bar chart showing the mean values and 95% confidence intervals and then add lower case letters to label the bars. Factor levels that do not have significantly different means will have the same letter label. In this example, cultivar 1 would be labelled 'a' and cultivars 2 and 3 would both be labelled 'b'. Using this method it is possible to have more than one letter. For example, if cultivar 1 was only different from cultivar 2, then cultivar 3, which is not significantly different from either of the other two, would be labelled with both 'a' and 'b'.

SPSS If you followed method 1 of 'Compare means' then 'one-way ANOVA', then you must select 'Post Hoc...' in the dialogue box before you choose 'OK'. This will offer you a choice of 18 *post hoc* tests, all currently unselected. Select either 'LSD' (least significant difference), 'S-N-K' (Student–Newman–Keuls), 'Sidak' or 'Bonferroni', then click on 'Continue' to return you to the main dialogue box.

The output from each test is similar. I have shown that from the LSD test below:

Post Hoc Tests

Multiple Comparisons

Dependent Variable: GRAIN_SZ

LSD

(I) CULTIVAR	(J) CULTIVAR	Mean Difference (I-J)	Std. Error	Sig.	95% Confidence Interval	
					Lower Bound	Upper Bound
1	2	-2.560*	.602	.001	-3.871	-1.249
	3	-2.520*	.602	.001	-3.831	-1.209
2	1	2.560*	.602	.001	1.249	3.871
	3	4.000E-02	.602	.948	-1.271	1.351
3	1	2.520*	.602	.001	1.209	3.831
	2	-4.000E-02	.602	.948	-1.351	1.271

*. The mean difference is significant at the .05 level.

This output table gives a line for each of the six pairwise comparisons (i.e. there is duplication as there are only three comparisons to be made). If a factor had four levels there would be 12 comparison lines. The 'Mean Difference (I-J)' gives the difference between the mean values for the two factor levels ('cultivars' in this case). Groups that are significantly different from each other are marked in this column with an asterisk. 'Std. Error' gives an estimate of the reliability of mean differences for the whole analysis. 'Sig.' gives the probability that the two groups have the same mean. The final two columns give confidence intervals for the difference. If both the lower and upper bounds have the same sign the two groups will be significantly different.

A comparison of this output with the Dunn–Sidák method shown above for the same data clearly shows that the Dunn–Sidák method is much more conservative.

MINITAB Before you click 'OK' to start a one-way test click the 'Comparisons ...' button. This takes you to a dialogue box with four different tests. Select the 'Fisher's individual error rate:' option and make sure that the number in the box is a 5, this is the percentage significance level so 5 translates as a critical *P*-value (alpha) of 0.05. This tests corresponds to the protected LSD test and should only be run if the *P*-value of the one-way ANOVA was less than 0.05.

When you run the test you get the following output after the standard ANOVA table:

```
                                   Individual 95% CIs For Mean
                                   Based on Pooled StDev
Level       N       Mean     StDev ----------+---------+---------+------
1           5     23.740     1.218 (-------*-------)
2           5     26.300     0.806                    (-------*-------)
3           5     26.260     0.764                    (-------*-------)
                                   ----------+---------+---------+------
Pooled StDev =     0.951                  24.0      25.2      26.4

Fisher's pairwise comparisons

    Family error rate = 0.116
Individual error rate = 0.0500

Critical value = 2.179

Intervals for (column level mean) - (row level mean)

                   1           2

          2    -3.8713
               -1.2487

          3    -3.8313     -1.2713
               -1.2087      1.3513
```

The graphical display of the mean and confidence interval for the means by groups shows quite clearly in this case that factor level '1' is very different from '2' and '3'. The 'Fisher's pairwise comparisons' section first confirms the error rate chosen — 0.05 in the 'Individual error rate' — and then gives the critical mean difference required for two factor levels (groups) to be deemed different at the 0.05 significance level. In the example group means have to be 2.179 apart to be deemed significantly different.

Finally comes a small table of the group comparisons showing the 95% confidence range for the mean difference between the groups. If both values are negative then the column level is significantly *lower* than the row level. If both are positive then the column level is significantly *higher* than the row level. If one is positive and one negative then the column and row levels are not significantly different. In the example level 1 'Premier' is significantly lower than both level 2 ('Super') and level 3 ('Dupa') while level 2 is not significantly different from level 3. If i is the number of groups then there will be $i(i-1)/2$ comparisons (i.e. three for three groups, six for four groups, and 45 for 10 groups) so the comparisons in this form become increasingly difficult to interpret.

Excel No direct methods are available. Although, if desperate, pairwise t-tests can be carried out *only* if the one-way ANOVA has already given a significant result and this will give the same result as an LSD test.

Kruskal–Wallis test

This test is the non-parametric equivalent of the one-way ANOVA and has a null hypothesis that all samples are taken from populations with the same median. It can be used to test any number of groups. However, unlike one-way ANOVA it does not make assumptions about homogeneity of variances or normal distributions. It is a typical 'rank' test, meaning that the raw data are converted into ranks before the test is carried out. The advantage of this is that it is ideal for situations where the highest value went off the scale or if extreme values are present, as these have a disproportionate influence on the results of parametric tests. This test *may* be used when there are only two samples, but the Mann–Whitney U test is more powerful for two samples and should be preferred.

This test is somewhat less powerful than one-way ANOVA, but you are less likely to find a significant result when there is no real difference (i.e. the probability of a type I error is decreased).

Example. I will be using the same data set as for the one-way ANOVA so the results from the two tests may be compared.

SPSS Under the 'Analyze' menu, choose 'Nonparametric Tests' and then select 'K Independent Samples...'. This will bring up a dialogue box for three non-parametric tests. By default the Kruskal–Wallis test box should be selected. Put the data ('grain_sz' in the example) into the 'Test Variable List' box by selecting it and then moving it across. Then select 'cultivar' as the 'Grouping Variable'. It will appear as 'cultivar(? ?)'. You need to click the 'Define Range...' button and then input '1' in 'Minimum' and '3' in 'Maximum' before clicking 'Continue'. This will return you to the first dialogue box. You will see that 'cultivar(? ?)' is now 'cultivar(1 3)'. Click 'OK' to run the test.

The output appears as follows:

NPar Tests

Kruskal-Wallis Test

Ranks

	CULTIVAR	N	Mean Rank
GRAIN_SZ	1	5	3.20
	2	5	10.40
	3	5	10.40
	Total	15	

Test Statistics[a,b]

	GRAIN_SZ
Chi-Square	8.655
df	2
Asymp. Sig.	.013

a. Kruskal Wallis Test

b. Grouping Variable: CULTIVAR

This confirms the test carried out is the Kruskal–Wallis. Next comes a table with confirmation that the variable 'grain_sz' was divided into groups by the variable 'cultivar'. Summary information about the three samples follows. The groups of 'cultivar' labelled '1', '2' and '3' each have an 'N' of 5 (cases or observations) giving 15 in total. The mean rank position of the groups (the smallest value of the 15 is given rank one and the largest rank 15), is also given. Even at this point it is clear that there is a difference between the cultivars as the mean rank of group '1' is very different to either group '2' or '3'.

Finally comes the output from the test in a small table. The first row gives a 'Chi-Square' value, the next an appropriate degree of freedom (there are three groups, giving 2 degrees of freedom) and then a P-value, labelled as 'Asymp. Sig.'.

The Kruskal–Wallis test can make adjustments for tied observations (two observations with exactly the same value). The output given by spss does make some correction for ties. A data set with a large number of tied observations will be much less likely to give a significant result. See below for *post hoc* tests.

MINITAB Input all the data into a single column and use a second column to label the cultivars with integers (as in the example). Label the columns appropri-ately. From the 'Stat' menu select 'Nonparametrics' then 'Kruskal–Wallis...'. In the dialogue box move the observed data into the 'Response:' box and the group codes into the 'Factor:' box. Click 'OK' to run the test.

(Or, if the command interface is enabled, type 'Krus c1 c2' (assuming the data are in c1 and the group codes in c2) at the MTB > prompt in the session window.)
You get the following output:

```
Kruskal-Wallis Test: Grain sz versus Cultivar

Kruskal-Wallis Test on Grain sz

Cultivar    N    Median    Ave Rank       Z
1           5    23.40          3.2    -2.94
2           5    26.40         10.4     1.47
3           5    26.00         10.4     1.47
Overall    15                   8.0

H = 8.64   DF = 2   P = 0.013
H = 8.66   DF = 2   P = 0.013 (adjusted for ties)
```

The output confirms the test used. Then it gives some summary information for each of the groups, giving their integer codes. The number of observations is 'N' here, and the median and mean rank are also given. The Z-value is used in the test calculation.

The last two lines give the test result. There are two versions depending on how they treat tied observations in the data. 'H' is a test statistic and 'DF' the degrees of freedom (one less than the number of groups). Finally the P-value, labelled 'P', is given. In the example it is far less than the critical 0.05 level and we reject the null hypothesis that the three groups have the same median. There is no *post hoc* test available to determine which group is different from which although this can be done by inspection of the summary statistics (and see below).

Excel There is no direct method, although it would be possible to carry out the test in the spreadsheet and obtain a significance value as the test statistic is distributed as chi-square.

Post hoc *testing: after the Kruskal–Wallis test*

The Kruskal–Wallis (KW) test has the same limitation as the one-way ANOVA in that a significant result just indicates that at least one pair of factor levels are significantly different from each other. The test does not indicate which pair, although inspection of the medians will show the extreme groups. However, there are two other pairwise combinations in a three factor level test and if there are four levels there is a total of six comparisons.

In one-way ANOVA a *post hoc* test will make all the pairwise comparisons and indicate which factor levels are different from which. There is not usually considered to be an equivalent for the KW test, but I suggest that, if the difference between each pair is an important thing to know, pairwise Mann–Whitney *U* tests (p. 101) should be carried out. This method has exactly the same logic as the LSD test if it is *only* applied when the KW test gives a significant result.

There are two independent ways of classifying the data

If for each observation you have two factors (different ways of subdividing the data into groups) and the factors are independent of each other (i.e. there is no way the level of one factor can be deemed to be affected by the level of another) then there are several tests for analysing the null hypothesis that all factor levels have the same mean. For example, you intend to record the beak widths of house sparrows in several different towns: your observations will be the measurement of beak width, one factor might be the sex of the bird and the other the town where it was collected.

In addition there may be a further null hypothesis that there is no interaction between the two factors under investigation. It is important to realize that interaction may only be investigated when there is more than one observation for each factor combination.

One observation for each factor combination (no replication)

There will be many circumstances where you wish to test the effect of two factors but are only able to take a very small number of observations. If you have only one observation for each factor combination then there are still tests you can perform to test the null hypothesis that each factor level has the same mean or median. The main difference in the interpretation of the results of these tests, when compared to tests with replication, is that there is no null hypothesis that there is no interaction between the two factors.

Example. A trial of six different blends of fertilizers (coded with the letters U to Z) has been carried out on linseed crops on four different farms (coded 1 to 4). Factor 1 is the fertilizer and factor 2 the farm. On each farm six fields were used in the trial so it was only possible to use each of the six fertilizer blends once on each farm. The crop yields of the linseed are given in the table below.

Farm	\multicolumn{6}{c}{Fertilizer blend}					
	U	V	W	X	Y	Z
1	1130	1125	1350	1375	1225	1235
2	1115	1120	1375	1200	1250	1200
3	1145	1170	1235	1175	1225	1155
4	1200	1230	1140	1325	1275	1215

Friedman test

The Friedman test is a non-parametric analogue of a two-way ANOVA (see below) that can only be used when there is a single observation for each factor combination (as in the data table above). The two null hypotheses are that the median values of each factor level are the same between columns and between rows. It does not make assumptions about the distribution of the data and can be used in any circumstance when the data are at least on an ordinal scale (can be put in a meaningful rank order). However, the test is rather conservative. Another problem is that in many packages the test has to be carried out twice: once to compare the rows and once for the columns.

SPSS The test compares columns only, so it has to be carried out twice. First, to test the null hypothesis that there is no difference between the columns (fertilizer blends in the example). Input the data into SPSS in exactly the same format as it is in the table of raw data. You can either label the columns with the code letters U–Z or leave them as the default 'var00001', etc.

From the 'Analyze' menu choose 'Nonparametric Tests' and then 'K related samples...'. In the dialogue box that appears move all the columns into the 'test variables' box. Make sure that 'Friedman' is checked and click 'OK'.

This output should appear:

Friedman Test

Ranks

	Mean Rank
U	1.50
V	2.50
W	4.50
X	4.88
Y	4.50
Z	3.13

Test Statistics[a]

N	4
Chi-Square	10.396
df	5
Asymp. Sig.	.065

a. Friedman Test

The first section of the output confirms the test used, then shows the mean rank position of the yield data for each of the 6 fertilizer blends (these are ranked 1 for the lowest on each farm and 6 for the highest). So the mean rank of blend 'U' is 1.5 on the four farms (it was lowest twice and second lowest twice).

The second section of the output ('Test Statistics') shows that there were four farms or cases ('N'), that the test output ('Chi-Square') was 10.396, there are 5 degrees of freedom, 'd.f.', and that the *P*-value ('Asymp. Sig.') is 0.065. This value is very close to the critical significance level of 0.05. So with such a conservative test we should seriously consider further trials.

You may wish to stop at this point or to continue depending on whether you are interested in the differences between farms or not.

To test the null hypothesis that there is no difference between farms the data need to be transposed so that each farm is in a different column. This is very easy in SPSS: select 'Data' then 'Transpose'. Move the column labels for each of the 6 fertilizer blends in the box labelled 'Variable(s)' and click 'OK'. The data should be rearranged so the rows become columns. They are given the default names 'var001', 'var002', etc.

Carry out the Friedman test as above, except do not use the variable 'case_lbl', and you get the following output:

Friedman Test

Ranks

	Mean Rank
VAR001	2.75
VAR002	2.17
VAR003	1.92
VAR004	3.17

Test Statistics[a]

N	6
Chi-Square	3.508
df	3
Asymp. Sig.	.320

a. Friedman Test

The details are as before. The most important number is the 'Asymp. Sig.' value of $P = 0.320$. We can therefore accept the null hypothesis that there is no difference between yields on different farms. In the example the variable names were left as the default SPSS names. These can be replaced easily by double-clicking on the column heading, or selecting the 'Variable View' and changing the 'Name' column.

MINITAB Input the observations into a single column. Use two further columns for the two grouping variables (factors) replacing the labels for factor levels with integers. Label the columns appropriately.

From the 'Stat' menu select 'Nonparametrics' and then 'Friedman...'. Put the observed data column into the 'Response:' box. Then put one of the factor columns into the 'Treatment:' box and the other into the 'Block:' box. If you are only really interested in one of the two factors then that should be in the 'Treatment:' box. Click 'OK' to run the test.

(*Or, assuming the observations are in column 1, and the two factors in columns 2 and 3, type* 'Friedman c1 c2 c3' *at the MTB > prompt in the session window.*)

You get the following output:

```
Friedman Test: yield versus fertilizer, farm

Friedman test for yield by fertiliz blocked by farm

S = 10.32   DF = 5   P = 0.067
S = 10.40   DF = 5   P = 0.065 (adjusted for ties)

                    Est      Sum of
fertiliz      N    Median    Ranks
1             4    1147.1     6.0
2             4    1161.3    10.0
3             4    1296.3    18.0
4             4    1255.8    19.5
5             4    1241.3    18.0
6             4    1193.3    12.5

Grand median  =    1215.8
```

The output confirms the test carried out, then the name of the column containing the data with the factor being tested ('fertiliz') in the example mentioned. Then come two lines with the test statistic 'S', the degrees of freedom 'DF' (one fewer than the number of factor levels) and the *P*-value associated with the test statistic. There are two versions of the test depending on how tied (equal) values are dealt with. The *P*-values will be similar unless there are a lot of tied observations in the observed data. Use the *P*-value in the second row (adjusted for ties). In this case the *P*-value is close to the critical 0.05 level so that although we accept the null hypothesis that there is no difference in median yield at different fertilizer levels we would consider further investigation. After the test comes some summary information about the different factor levels giving the number of observations ('N'), the estimated median yield for each factor level and then the mean rank position of the observations in each factor level.

This completes the investigation of one factor. To investigate the other factor: go to the 'Stat' menu, select 'Nonparametrics' then 'Friedman...' and swap the column names in the 'Block:' and 'Treatment:' boxes.

(*Or, if the command interface is enabled, type* 'Friedman c1 c3 c2' *at the MTB > prompt.*)

For the example data this gives:

```
Friedman Test: yield versus farm, fertilizer

Friedman test for yield by farm blocked by fertiliz

S = 3.45  DF = 3  P = 0.327
S = 3.51  DF = 3  P = 0.320 (adjusted for ties)

                    Est     Sum of
farm          N    Median   Ranks
1             6    1232.3    16.5
2             6    1213.6    13.0
3             6    1193.0    11.5
4             6    1255.5    19.0

Grand median  =    1223.6
```

The output is the same as for the first factor. In the example the *P*-value is well above 0.05 so we accept the null hypothesis that the farms have the same median yield.

There is no way to determine the significance of any interaction between the two factors.

Excel There is no direct method. It would be possible to calculate in Excel but this test requires a lot of data sorting.

Two-way ANOVA (without replication)

It may seem slightly odd that a statistical test that relies on the comparison of variation can be used when there is only one observation for each factor level combination but the test is perfectly valid. The two grouping variables may be set by the experimenter (e.g. different concentrations) or 'naturally' occurring (e.g. different sites). The assumptions of the test are: that the observed data are continuous; are approximately normally distributed; that the data would have about the same variance in each factor combination; and that the grouping variables have at least two levels each that can be coded into integers. Obviously with only one observation in each factor combination there is no scope for testing the data against a normal distribution. Therefore it is up to the tester to use common sense or previous knowledge about the data under investigation.

Example. The same example data set as for the Friedman test is used in the illustrations below. The data are crop yields that are rounded (therefore discontinuous) but there are clearly far more than 30 possible values so the assumption of continuous data can be accepted. The normal distribution might be more difficult as observations such as crop yields are often skewed to the right and would benefit from a log or square root transformation. However, in this case we accept that the raw data are suitable for the ANOVA.

The results of the ANOVA should be presented in table form:

Factor	d.f.	SS	MS	F-ratio	P-value
Farm	3	11071	3690	0.797	0.515
Fertilizer blend	5	59763	11953	2.58	0.071
Error	15	69479	4631		

Here d.f. is the degrees of freedom, SS is the sum of squares, MS is the mean square (MS = SS/d.f.) and F is the important statistic (F-ratio = MS factor/MS error). Error is often referred to as residual.

Note that there are 23 degrees of freedom in total in this example. There were 24 observations in the total data set so that is correct, the two main effects have six and four cases to give 5 and 3 degrees of freedom, leaving 15 in the error. It is always sensible to check the number of degrees of freedom in the output to see if it matches your expectations.

There is no possibility of examining the interaction between the two factors as there is no replication. The interaction is an important part of the power of analysis of variance where there is replication (see p. 133).

In the example the P-value associated with the fertilizer blends was quite close to the critical 0.05 level. In a report this could be highlighted in the text as follows: 'There was no significant difference in yield between farms ($F_{3,15} = 0.797$, $P > 0.1$) but there was an indication that yields varied between fertilizer blends ($F_{5,15} = 2.58$, $P = 0.071$)'. The subscripted numbers quoted with the F-ratios are the degrees of freedom.

SPSS Input the observed data into a single column. Use two separate columns to code the observations with integers to represent the factor levels. The integers do not have to be in sequence or start at 1. Label all the columns appropriately. In the example there will be a 'Yield' column, a 'Farm' column (containing the code numbers 1–4) and 'Fertiliz' column with number codes (1–6) replacing the letter codes from the original data. It is probably best to reduce the number of decimal places to zero at this point in the 'Variable View'. It is also possible to recode the numbers for the factor levels to more useful labels using the 'Values' column, although I have not done that here.

There are now at least two approaches to carrying out the test. They have different outputs so I will consider both. The first method assumes that both factors are 'fixed effects' (see p. 159), the second considers 'random effects'. Fixed effects would be testing the differences between particular fertilizers and farms while random effects would be testing the differences between any farms and any fertilizers. In this current example it is likely that fertilizer would be a fixed effect, while farm would be a random effect although it will make no difference to the significance values in this scenario.

Method 1. From the 'Analyze' menu select 'General Linear Model', then 'Univariate...'. In the dialogue box move the observations into the 'Dependent Variable' box and the two factors into the 'Fixed Factor(s):' box. Before you carry out the test you must click on the 'Model...' button and bring up the model box. Click on 'Custom', and then select both of the factors in the 'Factors & Covariates' list. From the drop-down list under 'Build Term(s)' select 'Main effects' and click on the move arrow to move the factors to the 'Model' box. In the example this gives 'Fertiliz' and 'Farm' only in the 'Model' box. If the line 'farm*fertiliz' appears, then remove it. If you do not do this step, the analysis will not provide any significance values. Click 'Continue' and once back to the original dialogue box click 'OK' to run the test.

You will get the following output:

Univariate Analysis of Variance

Between-Subjects Factors

		N
FERTILIZ	1	4
	2	4
	3	4
	4	4
	5	4
	6	4
FARM	1	6
	2	6
	3	6
	4	6

Tests of Between-Subjects Effects

Dependent Variable: YIELD

Source	Type III Sum of Squares	df	Mean Square	F	Sig.
Corrected Model	70833.333[a]	8	8854.167	1.912	.133
Intercept	35502337.5	1	35502337.50	7664.673	.000
FERTILIZ	59762.500	5	11952.500	2.580	.071
FARM	11070.833	3	3690.278	.797	.515
Error	69479.167	15	4631.944		
Total	35642650.0	24			
Corrected Total	140312.500	23			

a. R Squared = .505 (Adjusted R Squared = .241)

The output first confirms the test and then in a table 'Between-Subjects Factors' gives the number of observations in each level of the two factors. Then comes the ANOVA table itself labelled as 'Tests of Between-Subjects Effects'. The name of the dependent variable (i.e. the data column) is confirmed. There is more information than you actually need in the table (compare this table to the 'ideal' output table given in the description of the example). The important lines are for the factor variables ('Farm' and 'Fertiliz' in the example) and the 'Error' or residual. The 'd.f.' column gives the degrees of freedom, which is one less than the number of factor levels (in the example there were four farms so the factor 'Farm' has 3 d.f.). The (type III) sum of squares and mean square are given. The mean square is the sum of the squares/degrees of freedom. The last two columns give the important information. The 'F' column is the F-ratio (factor mean square/error mean square). The P-value is labelled 'Sig.' and is given in the last column. In the example the P-value for 'Farm' is well above the critical 0.05 level so we accept the null hypothesis that all the farms have the same mean yield. However, the P-value for 'Fertiliz' is 0.071, quite close to the critical level. We still accept the null hypothesis but should note that the P-value is close to the critical level in any report.

In this output it is best to totally ignore the lines labelled 'Corrected Model' (a combination of the two main effects), 'Total', 'Corrected Total' and 'Intercept' as not relevant to the two null hypotheses we are interested in.

Method 2. This time the test is carried out assuming that 'Farm' is a random factor and 'Fertiliz' is a fixed factor. From the 'Analyze' menu select 'General Linear Model' then 'Univariate...'. In the dialogue box move the observations into the 'Dependent Variable' box, 'Farm' in the 'Random Factor(s):' box and 'Fertiliz' into the 'Fixed Factor(s):' box. Before you carry out the test you must click on the 'Model...' button and bring up the model box. Click on 'Custom', and then select both of the factors in the 'Factors & Covariates' list. From the drop-down list under 'Build Term(s)' select 'Main effects' and click on the move arrow to move the factors to the 'Model' box. In the example this gives 'Fertiliz' and 'Farm(R)' only in the 'Model' box. If the line 'farm(R)*fertiliz' appears, then remove it. If you do not do this step, the analysis will not provide any significance values. Click 'Continue' and back in the main dialogue box you can either click 'OK' straight away or move first into the 'Options...' box where you can choose to display summary statistics for each of the variables. If you do this it can make interpretation of the output easier, although with no replication the distribution statistics cannot be calculated. Here I have selected 'Homogeneity tests'.

You get several tables of output. Here are the important ones:

Levene's Test of Equality of Error Variances[a]

Dependent Variable: YIELD

F	df1	df2	Sig.
.	23	0	.

Tests the null hypothesis that the error variance of the dependent variable is equal across groups.

a. Design: Intercept+FERTILIZ+FARM

Tests of Between-Subjects Effects

Dependent Variable: YIELD

Source		Type III Sum of Squares	df	Mean Square	F	Sig.
Intercept	Hypothesis	35502337.5	1	35502337.50	9620.505	.000
	Error	11070.833	3	3690.278[a]		
FERTILIZ	Hypothesis	59762.500	5	11952.500	2.580	.071
	Error	69479.167	15	4631.944[b]		
FARM	Hypothesis	11070.833	3	3690.278	.797	.515
	Error	69479.167	15	4631.944[b]		

a. MS(FARM)

b. MS(Error)

The first table, 'Levene's Test of Equality of Error Variances', would normally contain an estimate of the homogeneity of the variances in each factor combination. However, as there is only one observation in each, this cannot be calculated.

The second table, 'Tests of Between-Subjects Effects', contains the ANOVA information. The 'Dependent Variable' confirms the variable containing the data ('Yield' in the example). The table is different to the 'ideal' table shown above as it has some extra lines. The important lines are those with the names of the two factors. The five columns of the table give the SS (type III sum of squares), d.f. (degrees of freedom), MS (mean square), F-ratio (labelled as 'F') and a P-value (labelled 'Sig.') associated with the F-ratio and degrees of freedom. This is the value you compare with the critical level of 0.05. If the P-value > 0.05 then you must accept the null hypothesis that different levels of the factor have the same mean. In this example the P-value for 'Farm' is 0.515, well above 0.05, but that for 'Fertiliz' is, at 0.071, close to the critical level. This suggests that perhaps further investigation is required.

The line labelled as 'Intercept' should be ignored. Although it might be noted that the 'Sig.' is given as 0.000, when it should be reported as < 0.001.

The lines labelled 'Error' give the error or residual line for a conventional ANOVA table, although the line used as error for the 'intercept' is actually the line for the random effect (farm in this case). The error lines for the main effects should be reported in the ANOVA table.

MINITAB There are at least two ways of carrying out this test in MINITAB. I will only discuss one here as it gives the most useful output.

Input all the data into a single column and then use the next two columns for the factors replacing factor level names with integer codes. Label the columns appropriately. From the 'Stat' menu choose 'ANOVA', then 'General linear model …'. Move the data column into the 'Responses:' box and the two factor columns into the 'Model:' box. Click 'OK' to run the test.

(Or, *if the command interface is enabled, type* 'GLM c1 = c2 c3' *at the MTB > prompt in the session window*.)

If you used the example you get the following output:

General Linear Model: yield versus farm, fertilizer

```
Factor      Type Levels Values
farm        fixed    4  1 2 3 4
fertiliz    fixed    6  1 2 3 4 5 6

Analysis of Variance for yield, using Adjusted SS for Tests

Source     DF    Seq SS    Adj SS    Adj MS      F      P
farm        3     11071     11071      3690   0.80  0.515
fertiliz    5     59763     59763     11953   2.58  0.071
Error      15     69479     69479      4632
Total      23    140313

Unusual Observations for yield

Obs     yield       Fit    SE Fit  Residual   St Resid
 12   1140.00   1289.58     41.68   -149.58     -2.78R

R denotes an observation with a large standardized residual.
```

The output confirms that you have carried out a 'general linear model', a general term for a family of statistical tests that includes analysis of variance. Then it confirms the factors that have been used, how many levels there are for each factor and then the code numbers used for the factor levels (it pays to check these in case you made a mistake typing in the numbers!). Then comes the ANOVA table itself stating the name of the observation variable (yield in the example). The table has the usual columns except that it adds an additional SS column (with exactly the same numbers in as the usual SS column). 'Source' refers to the source of variation, 'DF' is degrees of freedom, 'SS' is sum of squares, 'MS' is mean square, 'F' is the *F*-ratio and 'P' the *P*-value. Compare the output to the ideal table I used in the description of the example. In the example the *P*-value for 'farm' is well above 0.05 so we accept the null hypothesis, but the *P*-value for 'fertiliz' is close to 0.05 so we may consider further investigation.

Finally comes a list of unusual observations. 'Obs' tells you which row number is unusual. These can often be typing errors so you should check the list against the original data set.

Excel There is no direct method unless you have installed the 'Data analysis toolpak' add-in.

1 Put the data into an Excel spreadsheet in exactly the format it is in the example with the appropriate labels or code letters as column and row labels.
2 Select the 'Tools' menu and select 'Data analysis....' (if this option does not appear you need to run Excel set-up and add the 'Data analysis toolpak' add-in). Select 'Anova: Two Factor Without Replication'.
3 A dialogue box will appear. If the cursor is flashing in the 'Input Range' box you can select the cells you wish to use for the analysis by clicking and dragging in the main sheet. If you select from cell A1 to the end of the data 'A1 : G5' should appear in the box. Alternatively you can just type in 'A1 : G5'. The tick box 'Labels' should be checked as these are included. Leave the 'Alpha' at 0.05 as this is the significance level which is chosen and $P < 0.05$ is the usual level.
4 At the bottom of the dialogue box is a section allowing you to determine where the output will appear. The default option is 'New Worksheet Ply' which means that the output will appear in a different sheet. If you want the output to appear on the same sheet as the data then you need to put a cell number in the 'Output range' box that will determine where the top left cell of the output will start.
5 Click on 'OK' and the following output will appear.

Anova: Two-Factor Without Replication

SUMMARY	Count	Sum	Average	Variance
1	6	7440	1240	11180
2	6	7260	1210	9230
3	6	7105	1184.167	1384.167
4	6	7385	1230.833	4054.167
U	4	4590	1147.5	1375
V	4	4645	1161.25	2606.25
W	4	5100	1275	11816.67
X	4	5075	1268.75	9322.917
Y	4	4975	1243.75	572.9167
Z	4	4805	1201.25	1156.25

ANOVA

Source of Variation	SS	df	MS	F	P-value	F crit
Rows	11070.83	3	3690.278	0.796702	0.51467	3.287383
Columns	59762.5	5	11952.5	2.58045	0.070826	2.901295
Error	69479.17	15	4631.944			
Total	140312.5	23				

The first section summarizes the data for each factor level of the two factors in turn. 'Count' is the number of observations and 'Average' the mean.

In the second section of the output there is a conventional ANOVA table. The 'Source of Variation' refers to the two factors. 'Rows' are farms and 'Columns' are fertilizer blends in this example. 'SS' is the sum of squares; 'd.f.' degrees of freedom (there were 24 observations in all giving 23 total degrees of freedom: there were four farms giving 3 d.f. and six fertilizer blends giving 5 d.f.); 'MS' is the mean square (= SS/d.f.) and 'F' is the ratio of the factor MS/error MS; and 'P-value' is the important value as it shows the probability that all factor levels (farms or fertilizer blends) have the same mean yield. In this case $P = 0.51467$ for 'Rows' (i.e. farms) and $P = 0.070826$ for 'Columns' (i.e. fertilizer) so we accept the null hypothesis that farms have the same mean yields and also accept the null hypothesis that the fertilizer blends have the same mean yield. However as the P-value was quite close to 0.05 we may consider another trial.

The final value 'F crit' is not usually quoted on an ANOVA table. It is the value of F required to achieve a $P = 0.05$ with the degrees of freedom in this particular test.

More than one observation for each factor combination (with replication)

The situation where you have two factors and more than one observation for each combination of factor levels is a very common one in biology. If there are two factors (ways of dividing the data into classes) then there are three hypotheses associated with the test: (i) that all levels of the first factor have the same mean; (ii) that all levels of the second factor have the same mean; and (iii) that there is no interaction between the two factors. If you have no interest in the third null hypothesis then you might consider two separate one-way analyses. However the extra power of the test that comes from investigation of the interaction is very great. Two tests are considered here: **two-way analysis of variance** and the **Scheirer–Ray–Hare test**.

Interaction

The concept of **interaction** warrants separate consideration. If the test gives a significant result for the interaction term it shows that the effect of the two factors in the test are not additive. This means that groups of observations assigned to levels of factor 1 do not respond in the same way to factor 2. For example, if you are measuring spiders from two locations you could have 'sex' as one factor and 'location' as the other (each spider can be assigned to one combination of sex and location). An analysis of variance gives a significant result for the factor 'sex' and for the factor 'location' as well as for the interaction term. This means that in some way the two sexes are responding differently in the two locations. The best way to interpret interaction is to plot out the means of the factor combinations roughly and inspect the graphs (Fig. 7.7).

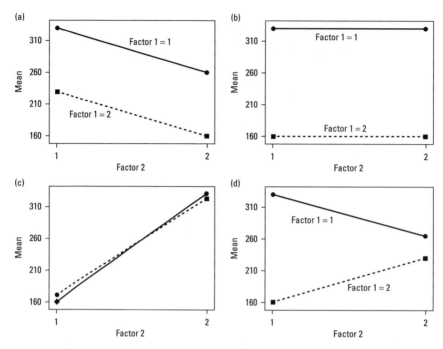

Fig. 7.7 In this figure there are four possible representations of the results of plotting mean values by factors after an analysis of variance. There are two factors labelled 'factor 1' and 'factor 2' here. Both have two levels labelled '1' and '2'. (a) A typical plot where both the main factors are significant but there is no interaction (lines are parallel). (b) Factor 1 is significant, factor 2 is not and there is no interaction. (c) Factor 2 is significant, factor 1 is not and there is no interaction. (d) Both factors are significant and there is also a significant interaction term — the lines are *not parallel* (i.e. the effect of factor 2 is different for the two groups of factor 1).

SPSS Interactions can be visualized in SPSS fairly simply. Once the analysis (either ANOVA or Scheirer–Ray–Hare) has been carried out, from the 'Graphs' menu select 'Line'. Then in the dialogue box select 'Multiple' and 'Summaries for groups of cases' before clicking on 'Define'. In the next dialogue select 'Other summary function' and move the observations column into the 'Variable:' box (it should default to 'MEAN(column)', if it does not click on 'Change Summary ...' and select mean). Then put one of the factors into the 'Category axis:' and the other into the 'Define lines by:' box (either way is fine, although using the factor with more group levels as the category axis is usually better). Click 'OK'. The graph produced is shown, slightly edited, in Fig. 7.8.

MINITAB There is a very simple way to visualize interactions in MINITAB. Once the analysis has been carried out following the instructions below, then a plot of the means for each factor and the effect of other factors can be produced. From the 'Stat' menu select 'ANOVA' and then 'Interaction Plot...' (Fig. 7.9). Move the

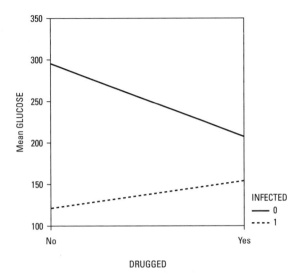

Fig. 7.8 In this example, produced in SPSS, the two lines are clearly not parallel, indicating that there is an interaction between the two factors. Here one factor is having an opposite effect on the two groups defined by the other factor. In the example the effect of the drug depends on whether the individual is infected or not.

column with the data into the 'Response:' box and the grouping variables into the 'Factors:' box. You can carry out this procedure after either a conventional ANOVA or a Scheirer–Ray–Hare test.

Excel Arrange the mean values for each factor combination in a table. Use row and column labels. For example:

		Subject unwell	
		0	1
Treated	0	295.66	207.50
	1	121.33	154.66

Select these cells. Then either use the 'Chart wizard' or from the 'Insert' menu select 'Chart' and 'On this sheet' before selecting the area where the chart will appear. In step 1 select the 'Line' graph. In step 2, 'Source data', the cells should be selected. The data should be in rows although often the plot is framed incorrectly making the data appear in columns. In step 3 extra details about the axes, etc. can be added. Decide where the chart will appear in step 4, then click 'Finish'. A chart something like that shown in Fig. 7.10 appears.

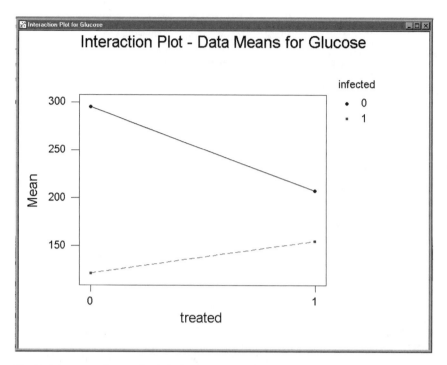

Fig. 7.9 This example uses the default 'Interaction Plot' option in MINITAB. There were two levels of two factors ('unwell' coded 0 or 1 and 'treated' coded 0 or 1). If there were no interaction between the two factors then the line joining the means would be parallel. In this example the lines are clearly not parallel and the treatment is clearly having an opposite effect on subjects (e.g. when 'unwell' is 0 there is a decrease in mean level with treatment, while when 'unwell' is 1 there is an increase).

Fig. 7.10 An interaction visualized in Excel. The means have to be calculated first and then plotted. The lines are clearly not parallel indicating that there is an interaction between the two factors.

Two-way ANOVA *(with replication)*

This is a very powerful statistical test with three null hypotheses: (i) the factor levels from the first main effect have the same mean; (ii) the factor levels from the second main effect have the same mean; and (iii) the two main effects do not interact (described above). It is suitable when there are two independent ways of assigning the observations into groups and there is more than one observation per factor combination. Assumptions are the same as with other ANOVA tests: the data are continuous (with more than 30 possible values); at least approximately normally distributed; and the variation is the same in each factor combination. The assumptions appear restrictive but two-way ANOVA is not very sensitive to slight violations of the assumptions. If you have data that clearly do not fit the assumptions you have five choices: do not do the test at all; try to transform the data to make it fit the assumptions; carry out a Scheirer–Ray–Hare test in lieu; use two separate one-way analyses instead; or use the test but be very cautious if the P-values are anywhere near the critical 0.05 level.

Note: 'balanced designs' have equal numbers of observations in each factor combination. The statistical calculations are simpler for balanced designs and the tests more powerful. Therefore when you design a controlled experiment it is always sensible to plan for equal numbers of observations from each factor combination (of course this is not always possible when making field collections or when plants die in laboratory experiments). However, as you are not calculating the tests by hand I will not consider 'unbalanced' and 'balanced' designs separately. Make sure that you avoid factor combinations with no observations at all. Unfortunately Excel will only work with balanced designs for two-way analyses.

Example. A biologist is investigating the effect of light on food intake in starlings. She sets up an experiment where the birds are placed in individual aviaries of identical size with controlled lighting and are given an excess of food. The birds are sexed and four male and four female birds are exposed to either 16 h or 8 h of light (e.g. long or short days). Each bird is monitored for 7 days and its total food intake in grams recorded. Each bird is only used once.

Day length	Sex							
	Female				Male			
Long (16 h)	78.1	75.5	76.3	81.2	69.5	72.1	73.2	71.1
Short (8 h)	82.4	80.9	83.0	88.2	72.3	73.3	70.0	72.9

There are 16 birds used in total. Four were exposed to each factor combination. Therefore, this is a balanced design.

The output from the statistical packages varies somewhat. It is usually not sensible to just put the whole output into a report. For a two-way ANOVA there is a standard way of displaying the output from the test. For the example, here the output would be presented in the following way:

Factor	d.f.	SS	MS	F-ratio	P-value
Day length	1	42.3	42.3	8.00	0.015
Sex	1	316.8	316.8	60.00	<0.001
Interaction	1	27.0	27.0	5.12	0.043
Error	12	63.4	5.28		

SPSS Input all the observations in a single column (Fig. 7.11). Use two further columns to input codes for the group labels associated with the two factors. The factor levels must be coded as integers. Label the columns appropriately (for the example, the factors have been labelled 'sex' and 'day_len'). I suggest that the number of decimal places is adjusted in the 'Variable View' so the factors have no decimal places and the data has one. The 'Values' column of the 'Variable View' can be used to label the factor levels as 'male' and 'female', or whatever is appropriate.

From the 'Analyze' menu select 'General Linear Model' and then 'Univariate …'. In the dialogue box move the observation column into the 'Dependent

Fig. 7.11 Two-way analysis of variance in SPSS. The two main effects have been defined as 'Fixed Factors' and the data has been moved to the 'Dependent Variable' box.

Variable:' box and the two coded factor columns into the 'Fixed Factor(s):' box. There is no need to use any of the options for a straightforward two-way design like this one. Click 'OK' to run the test.

Using the example you would get the following output:

Univariate Analysis of Variance

Between-Subjects Factors

		N
SEX	1	8
	2	8
DAY_LEN	1	8
	2	8

Tests of Between-Subjects Effects

Dependent Variable: DATA

Source	Type III Sum of Squares	df	Mean Square	F	Sig.
Corrected Model	386.130[a]	3	128.710	24.373	.000
Intercept	93025.000	1	93025.000	17615.591	.000
SEX	316.840	1	316.840	59.998	.000
DAY_LEN	42.250	1	42.250	8.001	.015
SEX * DAY_LEN	27.040	1	27.040	5.120	.043
Error	63.370	12	5.281		
Total	93474.500	16			
Corrected Total	449.500	15			

a. R Squared = .859 (Adjusted R Squared = .824)

The output confirms the test used. It then gives the number of observations for each level of both factors. Individuals will be double counted here (in the example there were 16 birds in total: 8 female and 8 male birds, 8 in short days and 8 in long).

Next comes the ANOVA table itself labelled as 'Tests of Between-Subjects Effects'. The name of the dependent variable (i.e. the data column) is confirmed. There is more information than you actually need in the table (compare this table to the 'ideal' output table given in the description of the example). The important lines are for the factor variables ('Sex' and 'Day_len' in the example), the interaction term ('Sex*day_len' in the example) and the 'error' or residual. The 'd.f.' column gives the degrees of freedom, which is one less than the number of factor levels (in the example there were two sexes and two day lengths so each has 1 d.f.). The degrees of freedom for the interaction term is the product of the degrees of freedom for the main effects (in this case both are one and 1×1 is 1). The (type III) sum of squares and mean square are given. The mean square is the sum of squares/degrees of freedom. The last two columns give the import-ant information. The 'F' column is the F-ratio (factor mean square/error mean square). The P-value is labelled 'Sig.' and is given in the last column.

In this output it is best to totally ignore the lines labelled 'Corrected Model' (a combination of the two main effects), 'Total', 'Corrected Total' and 'Intercept' as not relevant to the three null hypotheses we are interested in.

In the example the P-value for 'sex' is given as '0.000'; this is a highly signific-ant result but should always be reported as $P < 0.001$. We can be sure that the two sexes have different food intake levels although the test does not tell us which sex eats more — that must be done by inspection of the data. The P-value for the other factor ('Day_leng') is also less than 0.05 but the value of 0.015 indicates that the effect is not as strong as that for sex. Finally the interaction of the two factors also has a P-value less (albeit marginally less) than 0.05 meaning, in this example, that the two sexes respond significantly differently to day length in the amount of food they eat. The direction of this effect can only be revealed by inspection of the mean values for each group (see the section on interaction above).

Some of the options that might prove useful for a two-way ANOVA include 'Descriptive Statistics' and 'Homogeneity tests' in the options dialogue and a large range of *post hoc* tests (although with only two factor levels for each of the main effects a *post hoc* test just generates a warning message in SPSS). In the out-put below, a 'Profile plot' has been generated in the 'Plots' dialogue with the two factors as 'Horizontal Axis' and 'Separate Lines'.

Descriptive Statistics

Dependent Variable: DATA

SEX	DAY_LEN	Mean	Std. Deviation	N
female	1	77.775	2.529	4
	2	83.625	3.175	4
	Total	80.700	4.104	8
male	1	71.475	1.571	4
	2	72.125	1.475	4
	Total	71.800	1.453	8
Total	1	74.625	3.891	8
	2	77.875	6.560	8
	Total	76.250	5.474	16

Levene's Test of Equality of Error Variances[a]

Dependent Variable: DATA

F	df1	df2	Sig.
.888	3	12	.475

Tests the null hypothesis that the error variance of the dependent variable is equal across groups.

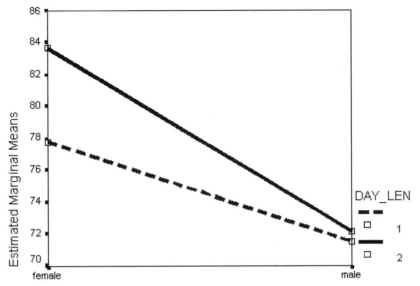

Estimated Marginal Means of DATA

The 'Descriptive Statistics' table usefully summarizes the full data set as well as each factor level and factor combination with means, standard deviations and number of observations. In the example I used 'Values' in the 'Variable View' to recode 'sex 1' as 'female' and 'sex 2' as 'male'. I did not recode the factor levels in 'Day_len' so they remain as '1' and '2'.

The 'Levene's Test of Equality of Error Variances' appears when the option 'Homogeneity tests' is selected.

MINITAB Input all the observations into a single column. Use separate columns to input the factor levels, coded as integers. Label the columns appropriately. There are then at least two ways of carrying out the test that give rather different outputs. I will consider both methods here.

Method 1. From the 'Stat' menu, select 'ANOVA' then 'General Linear Model …'. In the dialogue box put the column containing the data into the 'Responses:' box and then the two main effects (factors) into the 'Model:' box. To investigate interactions between the two factors the two column labels should be moved into the 'Model:' box again and an asterisk (*) put between them (in the example the 'Model:' box contained: sex 'day len' sex * 'day len' with day len appearing in quotes as the label is in two words). Click 'OK' to continue.

(*Or, if the command language has been enabled, type* 'glm c1 = c2 c3 c2*c3.' *at the MTB > prompt.*)

Using the example the following output appears in the 'Session' window:

```
General Linear Model

Factor       Type Levels Values
sex         fixed     2  1 2
day len     fixed     2  1 2

Analysis of Variance for data, using Adjusted SS for Tests

Source        DF    Seq SS     Adj SS     Adj MS      F       P
sex            1    316.84     316.84     316.84   60.00   0.000
day len        1     42.25      42.25      42.25    8.00   0.015
sex*day len    1     27.04      27.04      27.04    5.12   0.043
Error         12     63.37      63.37       5.28
Total         15    449.50

Unusual Observations for data

Obs      data       Fit   StDev Fit   Residual   St Resid
 12   88.2000   83.6250      1.1490     4.5750       2.30R

R denotes an observation with a large standardized residual.
```

First comes confirmation of the test (note that 'General Linear Model' includes ANOVA) and then the factors used, the number of levels (groups) and the integer codes assigned to the groups. (In the example there are obviously two levels for the factor 'sex' and they have been coded, alphabetically, as '1' for female and '2' for male).

The standard ANOVA table is next. It labels the factors and the interaction. The columns give degrees of freedom, 'DF'; two versions of the sums of squares, 'Seq SS' and 'Adj SS'; the mean square 'Adj MS', which is the sum of squares divided by the degrees of freedom; the F-ratio 'F', which is the mean square divided by the error mean square; and finally the P-value 'P'.

In the example the factor 'sex' has a P-value of 0.000 which should be reported as $P < 0.001$. We reject the null hypothesis that the two sexes have the same food intake confidently although we cannot tell from the table which sex has the higher intake. The factor 'day len' has a P-value well below 0.05 so we also reject the null hypothesis that day length has no effect of food intake. The interaction has a P-value slightly less than 0.05 so we reject the null hypothesis that day length has the same effect on both sexes but note that the result is not highly significant.

The next part of the output gives a list of unusual observations: it is always wise to check that these have been input correctly into the package. This part of the output can be suppressed by going to the 'Results...' dialogue before running the test and selecting the option 'Analysis of variance table'.

Method 2. From the 'Stat' menu, select 'ANOVA' then 'Two way...'. In the 'Two-way Analysis of Variance' box that appears move the column with the data to the 'Response:' box and the two main effects or factors ('sex' and 'day length' in the example) to the 'Row factor:' and 'Column factor:' boxes. Click 'Display means' for extra useful information in the output. Click 'OK'.

(Or, if the command language has been enabled, type 'twoway c1 c2 c3;' at the MTB > prompt and 'means c2 c3.' at the SUBC > prompt.)

The following output appears in the 'Session' window:

```
Two-way Analysis of Variance

Analysis of Variance for data
Source        DF        SS        MS        F        P
sex            1     316.84    316.84    60.00    0.000
day len        1      42.25     42.25     8.00    0.015
Interaction    1      27.04     27.04     5.12    0.043
Error         12      63.37      5.28
Total         15     449.50

                            Individual 95% CI
sex           Mean    -------+---------+---------+---------+----
1            80.70                                    (-----*-----)
2            71.80    (-----*-----)
                      -------+---------+---------+---------+----
                        72.00     75.00     78.00     81.00

                            Individual 95% CI
day len       Mean    ------+---------+---------+---------+-----
1            74.62    (--------*--------)
2            77.88                    (-------*-------)
                      ------+---------+---------+---------+-----
                        74.00     76.00     78.00     80.00
```

The ANOVA table is very much as it would appear in a publication or report with d.f., SS, MS, *F* and *P* clearly laid out (see the description of the data output above for more details). The 'two way . . .' route allows the differences between factor levels to be seen semigraphically. In the example the 95% confidence intervals for food taken for the two groups are very clearly different, reflecting the very low *P*-value. The intakes differ by day length, but the 95% confidence intervals overlap slightly reflecting the much higher, though still significant, *P*-value. It is also clear which groups are higher, with females (group 1) eating more than males (group 2) and those in short days (group 2) eating more than those in long days.

Excel This test is only possible in Excel if the design is balanced (i.e. each factor combination, long–female, short–female, etc., has the same number of observations). The data must be input so that one of the factors is separated into different columns and the other in sets of rows. The columns must be labelled with group labels and the first row of each group likewise.

The example was set up in Excel like this:

	Long	Short
Female	78.1	82.4
	75.5	80.9
	76.3	83
	81.2	88.2
Male	69.5	72.3
	72.1	73.3
	73.2	70
	71.1	72.9

From the 'Tools' menu select 'Data analysis…' then 'ANOVA: two factor with replication'. In the dialogue box that appears you must select the range of cells that contains not only the data but also the labels. This can be done either by dragging the mouse over the area of the relevant section of the spreadsheet or typing the top left and bottom right cell codes with a colon ':' between them (A1 : C9 in this case). Next, state how many observations there were for each factor combination (four in the example). Leave the 'Alpha:' as 0.05 (this determines the critical F-value which is quoted on the final table for comparison, i.e. the F-value required to reject the null hypothesis).

Finally, to select where the output will appear select one of the options in the bottom half of the box. Beware — the output covers a lot of cells.

The output is rather extensive:

Anova: Two-Factor With Replication						
SUMMARY	Long	Short	Total			
Female						
Count	4	4	8			
Sum	311.1	334.5	645.6			
Average	77.775	83.625	80.7			
Variance	6.395833	10.0825	16.84			
Male						
Count	4	4	8			
Sum	285.9	288.5	574.4			
Average	71.475	72.125	71.8			
Variance	2.469167	2.175833	2.111429			
Total						
Count	8	8				
Sum	597	623				
Average	74.625	77.875				
Variance	15.13929	43.03929				
ANOVA						
Source of Variation	*SS*	*df*	*MS*	*F*	*P-value*	*F crit*
Sample	316.84	1	316.84	59.99811	5.22E-06	4.747221
Columns	42.25	1	42.25	8.000631	0.015217	4.747221
Interaction	27.04	1	27.04	5.120404	0.042991	4.747221
Within	63.37	12	5.280833			
Total	449.5	15				

The first part of the output confirms the test carried out and then presents a summary table giving the number of observations, the sum, mean and variance for each factor combination, factor level and total. Some of this information is useful, especially when trying to interpret a significant result from the inter-action term. (Note that on older versions of Excel, the average and variance numbers given were incorrect for column and row totals.)

Then comes a rather standard ANOVA table that will require minimum editing before it can be included in a report. If you compare it with the 'ideal' table in the example you see how similar it is. The two factors are, unfortunately, only assigned as 'columns' and 'sample' although inspection of the original data in Excel will quickly determine which factor is which. The residual or error line is labelled as 'Within'. The columns of figures are sum of squares ('SS'), degrees of freedom ('d.f.'), mean square ('MS', where MS = SS/d.f.), F-ratio ('F', where $F = MS/MS_{error}$), the P-value and then the F-ratio that was required to reach a P-value of 0.05. The P-values are sometimes given in scientific notation so the '5.22E-06' in the example translates as 5.22×10^{-6} or 0.0000522, i.e. extremely highly significant.

In the example all three P-values are less than 0.05 so all three original null hypotheses are rejected.

Scheirer–Ray–Hare test

Non-parametric two-way ANOVA is usually deemed impossible in statistics books but the Scheirer–Ray–Hare (SRH) test is a non-parametric equivalent of a two-way ANOVA with replication. It is based on ranks so is suitable for any situation where the data can be put into order. It is quite easy (especially with fairly small data sets) to calculate a two-way ANOVA using ranks in an extension of the Kruskal–Wallis test called the Scheirer–Ray–Hare test. It makes few assumptions about the distribution of the data. However, the test is fairly recent and not yet widely available in statistical packages. For that reason it is a little awkward to carry out. This test is conservative and has a much lower power than the parametric ANOVA. There are further extensions to this test just becoming available to allow testing in multiway ANOVA. Unfortunately this is an area of considerable debate amongst statisticians and there are some workers who are unconvinced that the Kruskal–Wallis test can be extended to a two-way analysis. As this is, currently, the only two-way method available in lieu of parametric ANOVA, I suggest that it is used, but with some caution.

Example. I will use the same data of starlings grouped by sex and day length as for the two-way ANOVA. Unlike the standard ANOVA which rejects all three null hypotheses, the Scheirer–Ray–Hare test accepts two of the three null hypotheses associated with the experiment: there is no interaction between the two factors, and birds in different day lengths consume the same amount of food. This illustrates how much more conservative this test is compared to ANOVA.

1 Set out the data in exactly the same way as for parametric ANOVA. Sort the data (ascending) using the magnitude of the observations. From the 'Data' menu select 'Sort cases...'. In the dialogue box move the data column name into the 'Sort by:' box and select 'Ascending' (although it does not matter whether you choose ascending or descending). Click 'OK' and the rows will be shuffled into rank order.

2 Next you should assign ranks to replace the actual data (assign 1 to the smallest, etc.). You can either write over the original data observations or make a new column called 'rank'. Type the row number into this column. This is trivial when there are only 16 observations as in the example, but with hundreds of observations it would be better to import a list of ascending numbers from a spreadsheet.

3 Carry out a standard (parametric) ANOVA including the interaction. From the 'Analyze' menu select 'General Linear Model', then 'Univariate...'. In the dialogue box put the 'rank' column into the 'Dependent:' box and the two coded factor columns into the 'Fixed Factor(s)' box. Click 'OK' to run the test.

You will get this output if you use the example:

Between-Subjects Factors

		N
SEX	1	8
	2	8
DAY_LEN	1	8
	2	8

Tests of Between-Subjects Effects

Dependent Variable: RANK

Source	Type III Sum of Squares	df	Mean Square	F	Sig.
Corrected Model	285.000[a]	3	95.000	20.727	.000
Intercept	1156.000	1	1156.000	252.218	.000
SEX	256.000	1	256.000	55.855	.000
DAY_LEN	25.000	1	25.000	5.455	.038
SEX * DAY_LEN	4.000	1	4.000	.873	.369
Error	55.000	12	4.583		
Total	1496.000	16			
Corrected Total	340.000	15			

a. R Squared = .838 (Adjusted R Squared = .798)

4 This is not the end point for this test. You have to do the next bit by hand. Calculate the mean square value for the 'Total' row. This is not done in the SPSS output. It is the value of the 'Type III Sum of Squares' divided by the degrees of freedom. Call this value MS_{total}. (In the example MS_{total} is $1496/16 = 22.667$.)

The test statistic for the three null hypotheses is the value for the relevant sum of squares (SS) divided by MS_{total}. Calculate SS/MS_{total} for each factor and the interaction.

5 Using two new columns in the SPSS spread sheet, type in the values of SS/MS_{total} and the degrees of freedom using the same degrees of freedom as in a conventional ANOVA (i.e. one less than the number of factor levels for the two main factors and the degrees of freedom for the main factors multiplied together for the interaction). Label the columns, 'out' and 'd.f.'.

6 Then, to get the appropriate P-values for these values, from the 'Transform' menu select 'Compute...'. In the 'Compute Variable' window that appears, type the name of the column you want the result to go to in the 'Target Variable' box (e.g. 'pvalue'). Then in the 'Numeric expression:' box type '1–' (i.e. one minus) before selecting from the 'Functions' list 'CDF.CHISQ(q,d.f.)' and moving it into the 'Numeric expression:' box. Finally, replace the first question mark with the name of the variable where you input the SS/MS_{total} values and the second question mark with the name of the column with the degrees of freedom. Click 'OK' and the new column 'pvalue' will appear on the right-hand side of the spreadsheet with P-values from a chi-square table appropriate to the SS/MS_{total} values and the degrees of freedom. (You may need to increase the number of decimal places above 2 in the 'Variable View'.) (*Note*: you could look up the number in a chi-square table if you prefer.)

So for the example the SRH output could be displayed as:

	SS	SS/MS_{total}	d.f.	P-value
Day_leng (factor)	25	1.10	1	0.29427
Sex (factor)	256	11.29	1	0.00078
Day_leng sex (interaction)	4	0.176	1	0.67483

The results suggest that the only null hypothesis that must be rejected is that both sexes consume the same amount of food, as the P-value is < 0.001.

MINITAB There is no way of carrying out this test directly in MINITAB but with a little extra work the test is achievable.

1 Put the observations into a single column, then the factors in the next two columns with the factor levels coded as integers. Label the three columns appropriately. From the 'Manip' menu select 'Rank...'. Move the data observation column into the 'Rank data in:' box and in the 'Store ranks in:' type a new name (e.g. 'rank'). Click 'OK' and the rank position of the data appears in the new column with the smallest value ranked as 1.

2 Carry out a conventional two-way ANOVA using the 'rank' rather than the original observations. From the 'Stat' menu select 'ANOVA' then 'Two-way...'.

Put the rank column into the 'Response:' box and the two factors into the 'Row factor:' and 'Column factor:' boxes. Click 'OK'

(*Or, assuming the ranks are in c4 and the factors in c2 and c3, type* 'twow c4 c2 c3' *at the MTB > prompt in the session window.*)

The example will give the following output:

Two-way ANOVA: rank versus day, sex

```
Analysis of Variance for rank
Source        DF       SS       MS       F       P
day            1    25.00    25.00    5.45    0.038
sex            1   256.00   256.00   55.85    0.000
Interaction    1     4.00     4.00    0.87    0.369
Error         12    55.00     4.58
Total         15   340.00
```

---

3 Ignore the last two columns with the *F*- and *P*-values. Record the SS (sum of squares) values for the two factors and the interaction and calculate the total MS (SS_{total}/d.f.). Then calculate values for SS/MS_{total} either in MINITAB using 'Calc', then 'Calculator...' or by using a calculator. Type the calculated values into a single column in the spreadsheet (make sure to remember which value applies to each factor). Then use MINITAB to look up the values on a chi-square table: from the 'Calc' menu select 'Probability distributions' then 'Chi-square...'. Select the 'Cumulative probability' option. Then input the appropriate degrees of freedom (if the factors and interaction have different degrees of freedom then you will have to repeat this process for each different number of degrees of freedom); in the example there is 1 degree of freedom for all three values. In the 'Input column:' box put the name of the variable containing the calculated values and choose an empty column for the output 'Optional storage:'. Click 'OK'.

4 So far this has generated values for the probability of getting the number calculated *or lower*, whereas the number *or higher* is required for a *P*-value. So from the 'Calc' menu select 'Calculator...'. In the dialogue window put the name 'pvalue' in the 'Store result in variable:' box. Finally in the 'Expression:' box type '1–' (i.e. one minus), then the name of the column where the chi-square output was sent.

Using the example, this table shows the numbers used:

	SS	MS	SS/ MS_{total}	d.f.	Cumulative chi-square value	*P*-value
Day leng (factor)	25		1.10	1	0.705734	0.29427
Sex (factor)	256		11.29	1	0.999221	0.00078
Interaction	4		0.176	1	0.325166	0.67483
Total	340	22.67		15		

In the example the *P*-values for 'Day leng' and 'Interaction' are both well above 0.05 so the null hypothesis that the there is no difference between factor levels is accepted. However, the *P*-value for 'Sex' is < 0.001 indicating that the null hypothesis should be rejected and the two sexes have highly significantly different food intake.

Excel There is no direct way of carrying out this test in Excel although, providing the design is balanced (there is the same number of observations in each factor combination), it can be done using two-way ANOVA with only a few extra steps to rank the data and look up the significance of the test result.

1 First the raw data should be turned into ranks starting with the lowest value as 1, and then put into a table with row and column headings. So if the raw data were input as before, the observations would be in cells B2 to C9. To turn this into ranked information the 'RANK' function is used. In cell E2 type '= RANK(B2,B2 : C9,1)'. This gives the rank position of the datum in cell B2 in the set of 16 numbers. The default in Excel is to rank the largest value as 1. To rank the smallest value as 1, the option '1' is used in the 'RANK' function, as shown. Copy cell E2 to F2 and add labels. The example data set now becomes:

	Long	Short
Female	11	14
	9	12
	10	15
	13	16
Male	1	5
	4	8
	7	2
	3	6

2 Carry out a two-way ANOVA with replication exactly as described above. From the 'Tools' menu select 'Data analysis . . .' and then 'ANOVA: two-factor with replication'. Define the range to cover the cells with the ranked data and the group labels. Indicate how many observations there are for each factor combination, leave 'Alpha' as 0.05 and send the output to a clear area of the spreadsheet. Click 'OK'.

3 Ignore the output except for the ANOVA table at the bottom. Cut out the first four columns of the output and paste it into the top left corner of a new sheet. For the example the required output is:

Source of Variation	SS	df	MS
Sample	256	1	256
Columns	25	1	25
Interaction	4	1	4
Within	55	12	4.583333
Total	340	15	

Moving the output like this is not required by the package but it does remove distractions.

4 Calculate the total mean square MS_{total}. This is the total sum of squares (SS) divided by the total degrees of freedom (d.f.). If you have moved the table so that cell 'A1' contains the text 'Source of Variation' then just type in cell D7: '= b7/c7'. The statistic for the Scheirer–Ray–Hare test can now be calculated for the two main factors and the interaction, as it is SS/MS_{total}. In column 'e' type '= b2/d7' in row 2, '= b3/d7' in row 3 and '= b4/d7' in row 4. This gives three numbers that need to be looked up on a chi-square table — very easily done in Excel.

5 In column 'f' type 'P-value' in row 1 as a label; then in row 2 type '= CHIDIST(e2,c2)'. The P-value for the value in column 'e' and the degrees of freedom in column 'c' is given. Copy this into the two cells below (easily done in Excel by selecting cell 'f2' then clicking on the small black square in the bottom right corner of the cell and dragging it down to cover the next two cells).

The final table for the example data is as follows:

Source of Variation	SS	df	MS		P-value
Sample	25	1	25	11.29412	0.00078
Columns	256	1	256	1.102941	0.29362
Interaction	4	1	4	0.176471	0.67442
Within	55	12	4.583333		
Total	340	15	22.66667		

The P-values for 'Columns' (= day length in the example) and 'Interaction' are well above 0.05 so the null hypothesis that the factor levels have the same median is accepted. The P-value for 'Sample' (= sex) is well below 0.05. The null hypothesis is rejected and the alternative hypothesis that the two groups (sexes) have different food intakes is accepted. Which level is the higher must be determined by inspection.

There are more than two independent ways to classify the data

If each observation can be assigned to groups using more than two different factors (ways of classifying the data), each of the factors can be divided into groups and assigned integers and, most importantly, the factors are all fully independent of each other then the data are suitable for multifactorial testing.

The biggest danger is thinking that factors are independent when they are not. For example, if you sample from a variety of woodlands and use the wood number as a factor and then divide the samples into two groups based on whether they were in a northern or southern area, the woodland number would not be independent of the region (e.g. wood number 6 can only be in one or the other). If this is the case then a nested design is required rather than a 'fully factorial' one (see p. 159).

Multifactorial testing

If there are more than two factors and they are fully independent of each other then the design is said to be factorial. When there were just two factors there was scope for only one interaction term: factor A with factor B. As more factors are added to the analysis the number of interaction terms possible grows rapidly. For three factors there are four interaction terms ($A \times B$, $A \times C$, $B \times C$ and $A \times B \times C$); for four factors there are 11 terms.

Three-way ANOVA *(without replication)*

If time, space or resources are very limited then this design is quite common. If there are three independent ways of classifying the data into groups and there is only one observation for each factor combination then it is still possible to carry out an analysis of variance. Follow the instructions for three-way ANOVA with replication but be aware that there is no way of calculating the interaction between the three factors (this is used as the 'error' by the calculation). This means that only two-way interactions should be allowed in the model.

SPSS Follow the instruction for three-way ANOVA with replication but if you have no replication (i.e. there is only one observation for each factor combination) when you have arranged the factors and the dependent variables you must first go to the 'Model...' submenu. Click on 'Custom'. Then highlight the three factor names and under 'Build terms' select 'Main effects' from the menu before moving the factors across to the 'Model:' box. Then repeat with 'All 2-way' selected. In the example this will give six lines in the 'Model:' box — each effect and three pairwise interactions. Click 'Continue' to return to the 'Univariate' dialogue and 'OK' to run.

MINITAB Follow the instructions for three-way ANOVA with replication but when specifying the 'Model:' the three-way interaction (e.g. c2*c3*c4) should be omitted. If column names are not used and the grouping variables are in columns 2, 3 and 4 this would give a model statement of 'c2 c3 c4 c2*c3 c2*c4 c3*c4'.

Excel Only one-way and balanced two-way ANOVAs are possible in Excel.

Three-way ANOVA *(with replication)*

If there are three fully independent ways of dividing the data into groups (e.g. site of collection, species and sex) and there is more than one observation for each factor combination then this design is appropriate. As with all analysis of variance the test assumes that the data are continuous, approximately normally distributed and that the variance is approximately equal for each factor level. Unfortunately there is no non-parametric equivalent to use if the assumptions are not met.

Example. A group of agricultural ecologists are interested in the effect of rabbit grazing on plant communities near to warrens. The observations are carried out on two sites on different soils. At each site there are two areas for study, one is 10 m from an active warren and one is 100 m from a warren. In each study area six quadrats are positioned at random. Two quadrats are left untouched as a control, two have caged exclosures placed around them that totally exclude rabbits and two are 'procedural controls' where the digging for constructing an exclosure is carried out but the rabbits are not excluded with fences. A procedural control is required in this experiment as otherwise any effects could be attributed either to the disturbance from construction or the exclusion of rabbits. Procedural controls are often overlooked when designing experiments (see p. 27).

The data are the standing crop of grass collected 100 days after the exclosures were placed. There are three independent factors: site (two levels, coded 1 and 2); distance from warren (two distances, 10 m and 100 m, coded 1 and 2); exclosure type (three levels, coded: 0, control; 1, procedural control; 2, exclosure). There are two observations for each factor combination with a total of 24.

Here is the data in table format:

		Distance from warren					
		1			2		
		Exclosure type					
		0	1	2	0	1	2
Site	1	112	115	187	141	121	189
		116	102	175	101	157	186
	2	121	145	198	135	141	208
		138	124	168	129	133	206

The output from statistical packages often has more information than is required. The ANOVA table included in a report or publication should be presented as follows:

Source of variation	d.f.	SS	MS	F	P
Distance	1	1218.4	1218.4	6.44	0.026*
Exclosure	2	6841.4	3420.7	18.07	<0.001***
Site	1	590.0	590.0	3.12	0.103
Distance × exclosure	2	327.0	163.5	0.864	0.446
Distance × site	1	1.0	1.0	0.006	0.942
Exclosure × site	2	58.3	29.2	0.154	0.859
Distance × exclosure × site	2	282.3	141.2	0.746	0.495
Error	12	2271.5	189.3		
Total	23				

This sort of table should always be accompanied with a table caption that explains the test used in more detail without making the reader refer to the main text. The asterisks used to indicate levels of significance should also be translated: $* = P < 0.05$, etc.

This example assumes that the factors are all fixed; that is the factor levels are meaningful, rather than arbitrary, labels (see p. 159 for a discussion of fixed and random effects). In this example there is no doubt that 'Distance' and 'Exclosure' are fixed effects as the groups have clear meanings. The factor 'Site' is one that might be either fixed or random. If treated as fixed then the analysis considers differences between two specified sites. If treated as random then the difference considered is that between any two sites and their location is unimportant.

SPSS The test is essentially the same as for two-way ANOVA. Arrange the data so that there is one column for the observations and one for each of the three factors with the group labels coded as integers. The group labels can be decoded in the 'Variable View' using the 'Values' column, although I have not done this in the example. It will be assumed that the three factors are all fixed effects.

From the 'Analyze' menu select 'General Linear Model' then 'Univariate...'. Put the observations into the 'Dependent:' box and the three factors into the 'Fixed Factor(s):' box. Define the range (highest and lowest coded integer) for each factor. Click 'OK' to run the test.

Using the example you get the following output:

Univariate Analysis of Variance

Between-Subjects Factors

		N
DISTANCE	1	12
	2	12
EXCLOSE	0	8
	1	8
	2	8
SITE	1	12
	2	12

Tests of Between-Subjects Effects

Dependent Variable: DATA

Source	Type III Sum of Squares	df	Mean Square	F	Sig.
Corrected Model	23565.333[a]	11	2142.303	10.349	.000
Intercept	524512.667	1	524512.667	2533.878	.000
DISTANCE	888.167	1	888.167	4.291	.061
EXCLOSE	21085.083	2	10542.542	50.930	.000
SITE	864.000	1	864.000	4.174	.064
DISTANCE * EXCLOSE	166.583	2	83.292	.402	.677
DISTANCE * SITE	37.500	1	37.500	.181	.678
EXCLOSE * SITE	6.250	2	3.125	.015	.985
DISTANCE * EXCLOSE * SITE	517.750	2	258.875	1.251	.321
Error	2484.000	12	207.000		
Total	550562.000	24			
Corrected Total	26049.333	23			

a. R Squared = .905 (Adjusted R Squared = .817)

The output confirms the test and then gives the number of observations in each level of the three main effects. It is worth checking this to make sure that the data has been input correctly. Note that each datum will be counted three times in this table.

Next comes the ANOVA table. First the name of the dependent variable is confirmed. Then comes an ANOVA table with some extra lines. The most important lines are for the factor variables, called 'Main Effects' in SPSS ('Distance', 'Exclose' and 'Site' in the example) and their interactions (three possible two-way interactions and one three-way interaction). The 'd.f.' column gives the degrees of freedom, which is one less than the number of factor levels for main effects and is calculated by multiplication for interactions (e.g. for 'distance * exclose' there are $2 \times 1 = 2$ d.f.). The sum of squares and mean square are given (mean square is the sum of squares/degrees of freedom). The last two columns give the important information. The 'F' column is the F-ratio (factor mean square/error mean square). The *P*-value, labelled 'Sig.', is in the last column. In the example the *P*-values for both 'Site' and 'Distance' are slightly above the critical 0.05 level so we accept the null hypotheses that there is no difference between sites or between distances. However, the *P*-value for 'Exclose' is given as 0.000 (i.e. < 0.001), much less than 0.05, so we reject the null hypothesis and accept the alternative hypothesis that exclosure affects the grazing of rabbits. None of the interaction terms has a *P*-value even close to 0.05.

In this output it is best to totally ignore the lines labelled 'Corrected Model', 'Corrected Total' and 'Intercept'.

POST HOC TESTING IN SPSS The ANOVA has indicated that one of the main effects is significant. If there were only two factor levels, we could easily determine which group was higher by inspecting the means of the data. However, as there are three factor levels for the significant effect a *post hoc* test is required. Repeat the method as above but in the 'Univariate' box click on 'Post hoc'. This brings up a dialogue box with a bewildering array of options. First move the required factor from the 'Factor(s):' box to the 'Post Hoc Tests for:' box. This will activate all the options for *post hoc* tests. All will tend to give the same results, but some will be more conservative than others. I suggest the 'Sidak' method, 'SNK' or 'LSD' (least conservative) tests, although you should only conduct an LSD test if the factor is significant.

This is the output from three *post hoc* tests using the example data:

Post Hoc Tests

EXCLOSE

Multiple Comparisons

Dependent Variable: DATA

	(I) EXCLOSE	(J) EXCLOSE	Mean Difference (I-J)	Std. Error	Sig.	95% Confidence Interval Lower Bound	95% Confidence Interval Upper Bound
LSD	0	1	-5.63	7.19	.449	-21.30	10.05
		2	-65.50*	7.19	.000	-81.17	-49.83
	1	0	5.63	7.19	.449	-10.05	21.30
		2	-59.88*	7.19	.000	-75.55	-44.20
	2	0	65.50*	7.19	.000	49.83	81.17
		1	59.88*	7.19	.000	44.20	75.55
Sidak	0	1	-5.63	7.19	.833	-25.55	14.30
		2	-65.50*	7.19	.000	-85.43	-45.57
	1	0	5.63	7.19	.833	-14.30	25.55
		2	-59.88*	7.19	.000	-79.80	-39.95
	2	0	65.50*	7.19	.000	45.57	85.43
		1	59.88*	7.19	.000	39.95	79.80

Based on observed means.

*. The mean difference is significant at the .05 level.

Homogeneous Subsets

DATA

			Subset	
	EXCLOSE	N	1	2
Student-Newman-Keuls [a,b] 0		8	124.13	
	1	8	129.75	
	2	8		189.63
	Sig.		.449	1.000

Means for groups in homogeneous subsets are displayed.
Based on Type III Sum of Squares
The error term is Mean Square(Error) = 207.000.

   a. Uses Harmonic Mean Sample Size = 8.000.

   b. Alpha = .05.

There are two tables. The first gives the LSD and Sidák method results. There are six lines for each method giving each pairwise comparison. This means that each pair appears on the list twice (i.e. as '0'–'1' and '1'–'0'). The table gives the difference between the mean observations from the two factor levels, a measure of the standard error of the differences and, most importantly, a 'Sig.' column. In the example, for both the LSD and Sidák methods there is no significant difference between groups 0 and 1, but there is a highly significant difference between groups 0 and 2 and between 1 and 2. This indicates that areas surrounded by exclosure fences have, perhaps unsurprisingly, a much higher mean grass weight.

The second table gives the same information in a different format as the result of the SNK test. This *post hoc* test puts groups 0 and 1 in a 'homogeneous subset' that is different from group 2. This table also, usefully, gives the mean values of the observations in each group.

MINITAB Put the observations into a single column. Use three further columns for the group labels of the three factors (coded as integers). Label the columns appropriately. From the 'Stat' menu select 'ANOVA' then 'General Linear Model...'. Move the observation column into the 'Responses:' box and the three factor columns into the 'Model:' box. Unfortunately if interactions are to be investigated these need to be added separately. So, assuming the factors are in columns 2, 3 and 4 then the following needs to be added to the model box: 'c2*c3 c2*c4 c3*c4 c2*c3*c4'. The names of the factors could be used instead of column codes. (If there is no replication (i.e. only one observation for each factor combination) then the final c2*c3*c4 term should be omitted.)

The following output appears:

General Linear Model: data versus site, dist, exclosure

```
Factor      Type Levels Values
site        fixed    2  1 2
dist        fixed    2  1 2
exclosur    fixed    3  0 1 2

Analysis of Variance for data, using Adjusted SS for Tests

Source              DF    Seq SS    Adj SS    Adj MS      F      P
site                 1     864.0     864.0     864.0   4.17  0.064
dist                 1     888.2     888.2     888.2   4.29  0.061
exclosur             2   21085.1   21085.1   10542.5  50.93  0.000
site*dist            1      37.5      37.5      37.5   0.18  0.678
site*exclosur        2       6.3       6.3       3.1   0.02  0.985
dist*exclosur        2     166.6     166.6      83.3   0.40  0.677
site*dist*exclosur   2     517.7     517.7     258.9   1.25  0.321
Error               12    2484.0    2484.0     207.0
Total               23   26049.3
```

This is essentially the same as the output for the two-way ANOVA only with several more lines. I always advise checking the first part of the output confirming the factor names, number of groups and the integer codes. The ANOVA table contains lines for each of the factors, then the interactions and finally the error and total. The P-values (labelled 'P') are less then 0.05 for only one of the factors and none of the interactions. The null hypotheses that grass weight is the same for all levels of these two factors is rejected and an alternative hypothesis is accepted. However, in this case the two non-significant factors both have P-values near to 0.05 and further analysis should be considered.

Visualization of which factor level is higher than which is simple to achieve in MINITAB. From the 'Stat' menu select 'ANOVA' and then 'Interactions plot...'. Put the factor columns into the 'Factors:' box (no need for the interaction terms this time) and the raw data in the 'Raw response data in:' box. Click 'OK' and a set of graphs showing the mean values of the raw data for every factor combination appears (Fig. 7.12). It is clear that the data from level 2 of the exclosure factor is much higher than the other two levels of this factor. (See the interaction section for an explanation of how these graphs can be interpreted, p. 133.)

Excel There is no direct way to carry out this test.

Multiway ANOVA

If there are more than three ways of dividing the data into groups and each of the classifications is independent of the others then ANOVA may be carried out. These multifactorial designs become increasingly difficult to interpret as there is an explosive increase in the number of interaction terms as the number of factors goes up. Furthermore, unless an experiment is designed to be fully factorial there are likely to be combinations of factors where there are no observations.

Finally, once there are many factors the chance that they are all independent of each other diminishes. Each factor should be considered in turn — is it

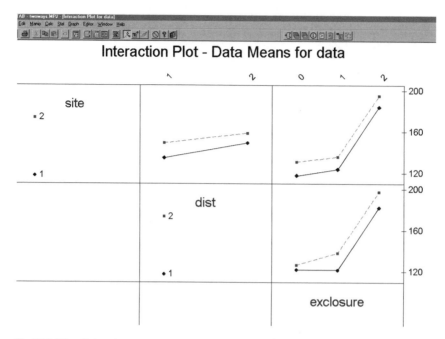

Fig. 7.12 Visualizing three-way ANOVA in MINITAB using the 'Interaction Plot'. The factor 'exclosure' has clearly higher means for level 2 when divided by 'dist' or 'site'.

independent of all other factors? Can two factors be sensibly combined in some way? Is one factor nested within another (see next section)? Is one factor more appropriately investigated as a covariate (see the section on analysis of covariance, p. 195)?

Not all classifications are independent

A very large number of experiments that are treated as if all the factors are independent prove to be, on further investigation, nothing of the sort. Therefore it is important that the independence of the factors in the analysis is carefully considered.

Non-independent factors. Some ways of classifying data are always going to be fixed and are therefore main factors in an analysis. For example, a commonly used factor in biology is 'sex' — there are two sexes (although in some species there are intermediate 'intersexes' or hermaphrodites) and once an individual is labelled as either male or female that classification is fixed. Other factors are not so clearly fixed: if a set of observations is divided into two equal-sized groups by weight then an individual may find itself in the heavy group at the start but later moved into the light group. This type of factor is not really fixed although the label 'heavy' really means something about the observation. Consider a study on blood pressure in humans using the two factors I have described here: 'sex' and 'weight'. Males are heavier than females and therefore there will be more males in the heavy group; therefore *the two factors are not independent of each other*. This type of independence can be checked using a chi-square test of association considered in the next chapter (p. 165). If the factors are not independent then two-way ANOVA is inappropriate. One solution is to apply an analysis of covariance (ANCOVA) using 'sex' as a fixed factor and 'weight' as a covariate. This analysis is considered in the next chapter (p. 195).

Nested factors. Nested factors are never 'fixed', meaningful factors, they are always factors numbered for convenience where the number given does not really mean anything. For example in a greenhouse experiment there may be plants in numbered pots with four pots per numbered tray and three trays per bench with two benches in each of two greenhouses. At each level — 'pot', 'tray', 'bench' and 'greenhouse' — the numbers are only used for convenience. These are called 'random' factors. In this example, the factor 'pot' is said to be nested within 'tray', the 'trays' nested within 'bench' and 'bench' within 'greenhouse'. It is important that the numbering system reflects this nesting: pots should be labelled 1–4 in *each* tray and trays labelled 1–3 in each greenhouse. *Nested factors are always random but not all random factors need to be nested* (e.g. the factor 'site' in the three-way ANOVA example might be a random factor, but it is not nested). *Post hoc* tests may not be carried out on nested factors.

Random or fixed factors. I have touched on the difference between 'fixed' or meaningful factors and 'random' or convenience factors above. There is, however, not a strict distinction between the two types of factors and the same factor can be treated as 'fixed' or 'random' in different tests. For example, in an analysis of behaviour in *Drosophila melanogaster* mutants the mutants used could be either a 'fixed' or 'random' factor. If the mutant is treated as 'fixed' (so that *vestigial* is coded as 1 and *white-eye* as 2) then this implies that the significance level of any difference is due to the characteristics of the particular named mutants. However, if the factor is 'random' then any significance implies differences between *any* two randomly selected mutants.

This section considers designs with a simple hierarchy of nested factors. More complicated designs, where it is appropriate to test both nested and factorial designs combined, are surprisingly common. These designs are extremely common in biology but they are often very difficult to achieve in the statistical packages.

Nested or hierarchical designs

If the experiment or sampling design has only one 'random' factor at the top of the hierarchy and all other factors are 'nested' within it then the design is said to be a pure nested one. If the top level is 'fixed' and there are other factors nested within it the design is often called a 'mixed model'. There is no limit to the number of factors that can be nested in this way. The simplest case of a single fixed factor and one nested is considered first then a more complex one.

Two-level nested design ANOVA

This is an analysis with two factors. The standard factor in this type of design may be 'fixed' or 'random'. There should then be a second, 'random' factor nested within the standard factor.

Example. In an experiment on cholesterol level in the blood of mice, two levels of fat intake are fixed by the researchers and coded as level '0' and '1'. For each level of 'intake' there are three populations of mice in separate cages and from each of these cages three individual mice are selected at random for blood testing. The six populations of mice are assigned to the factor 'cage', which is a 'random' factor nested within 'intake', with the three cages in each group coded '1', '2' and '3' (clearly these labels have no real meaning and are only used for convenience). There are six cages in all but they must not be coded as 1 to 6.

The data collected are as follows:

	Intake					
	0			1		
	Cage					
Mouse	1	2	3	1	2	3
1	55.6	62.5	58.9	85.6	68.5	65.6
2	62.5	68.3	54.2	98.2	69.3	71.0
3	68.2	58.2	63.5	75.1	88.2	78.3

Note: a common error made when analysing this type of data is that 'mouse' is treated as a factor. This might seem sensible as the mice are labelled 1–3. However, it is inappropriate as there is only one observation per mouse; ANOVA would have no variation to work with if it was used as a factor. Therefore 'mouse' is actually the level of replication within the factor 'cage'. If there were two observations per mouse then 'mouse' could be a further 'random' factor that would be nested within 'cage'.

Using the example data the following ANOVA table should be used in a report:

Source of variation	d.f.	Sum of squares	Mean square	F-ratio	P-value
Intake	1	1215.24	1215.24	18.90	0.001**
Cage within intake	4	377.42	94.35	1.47	0.272
Error	12	771.68	64.31		
Total	17				

SPSS Nested analyses are achievable, but slightly awkward, in SPSS. Put the observations in a single column. Use two further columns for the factors with the levels coded as integers. From the 'Analyze' menu select 'General Linear Model', then 'Univariate...'. Put the observations into the 'Dependent variable:' box and the fixed factor in the 'Fixed Factor(s):' box ('intake' in the example is a fixed effect). Put the nested factor in the 'Random Factor(s):' box. (A detour via the 'Options...' button or the 'Plots...' button can be useful as a means by which factor levels can be requested, allowing an easy comparison of groups.)

Now the awkward bit...click on 'Paste'. This brings up the command line 'Syntax' window of SPSS. It will probably be called 'Syntax1'. For the example it contains the following:

```
UNIANOVA
  data BY intake cage
  /RANDOM = cage
  /METHOD = SSTYPE(3)
  /INTERCEPT = INCLUDE
  /CRITERIA = ALPHA(.05)
  /DESIGN = intake cage intake*cage
```

To carry out a nested ANOVA the '/design' line must be altered to specifically request it. The nested factor must be specified as being within the main factor. The nested factor is listed before the factor it is nested within and, rather coun-terintuitively, the fixed factor has to be in brackets. In the example the last line becomes '/DESIGN cage(intake) intake.'. To run this design from the 'Syntax' window click on the 'play' button (an icon of a small black triangle pointing to the right).

The following output appears:

Univariate Analysis of Variance

Between-Subjects Factors

		N
INTAKE	0	9
	1	9
CAGE	1	6
	2	6
	3	6

Tests of Between-Subjects Effects

Dependent Variable: DATA

Source		Type III Sum of Squares	df	Mean Square	F	Sig.
Intercept	Hypothesis	87041.827	1	87041.827	922.498	.000
	Error	377.418	4	94.354[a]		
INTAKE	Hypothesis	1215.245	1	1215.245	12.880	.023
	Error	377.418	4	94.354[a]		
CAGE(INTAKE)	Hypothesis	377.418	4	94.354	1.467	.272
	Error	771.680	12	64.307[b]		

a. MS(CAGE(INTAKE))

b. MS(Error)

Expected Mean Squares[a,b]

	Variance Component		
Source	Var(CAGE (INTAKE))	Var(Error)	Quadratic Term
Intercept	3.000	1.000	Intercept, INTAKE
INTAKE	3.000	1.000	INTAKE
CAGE(INTAKE)	3.000	1.000	
Error	.000	1.000	

First comes a table confirming the number of observations for each factor level. This is worth checking to confirm it matches your expectations. It can often highlight an error in the data entry.

The ANOVA table confirms the name of the dependent variable. Then comes a conventional ANOVA table with slightly more information than is needed. The important lines are those labelled with the names of the main effect and nested factor and the bottom error (or residual) line. In the example the P-values (labelled 'Sig.') show that there is no effect of 'Cage(intake)' (i.e. 'cage' nested within 'intake') but there is a significant effect of 'Intake' as P is less than 0.05. The degrees of freedom for the nested factors are slightly odd as they have one less than the number of levels for *each* level of the main factor (in the example this is the number of cages − 1 for each intake or $(3 − 1) \times 2 = 4$).

The final table gives the expected mean squares. These are very important when calculating ANOVA by hand, but are not useful here.

A possible complication: note that in this case SPSS has generated a different *F*-ratio to that shown in the table above. This is because the calculation of *F*-ratios in nested ANOVA is something of an art. In this case SPSS has used the mean square of the nested factor as the denominator for the *F*-ratio while the table above and MINITAB have used the mean square of the error instead. There is a convention that the mean square of the nested factor is used as the denominator *whenever the factor is significant* but in other circumstances there is debate over the best denominator to use. So, if the nested factor has a *P*-value less than 0.05 it is customary to use the mean square of this factor as the denominator to calculate the *F*-ratio of the main factor rather than the error mean square. SPSS VERSION 10 appears to conform to this convention, but it would be worth checking the *F*-ratios carefully if you have a different version of the package. With more complicated designs selection of the correct denominators can become a rather long-winded process.

MINITAB Put the observations in a single column. Use two further columns for the factors with the levels coded as integers. From the 'Stat' menu, select 'ANOVA' and then 'General Linear Model…'. Move the observation column into the 'Responses:' box. Then put the main factor and the nested factor ('intake' and 'cage' in the example) into the 'Model:' box. To show that the nested factor is nested it must be followed by the name of the main factor in brackets (in the example the 'Model:' box reads: 'intake cage(intake)'). Click 'OK'. (A detour via the 'Options…' button can be useful as a means by which factor levels can be requested, allowing comparison of groups.) You might note that there is a 'Fully Nested ANOVA' option within the ANOVA model, but I find that this is much less controllable than using 'General Linear Model' in MINITAB.

(Or, if the command interface is enabled, and assuming data in c1, main factor in c2 and nested factor in c3, type 'GLM c1 = c2 c3(c2)' at the MTB > prompt in the session window.)

You get the following output from the example:

```
General Linear Model: blood versus Intake, Cage

Factor          Type Levels Values
Intake          fixed     2  0 1
Cage(Intake)    fixed     6  1 2 3 1 2 3

Analysis of Variance for blood, using Adjusted SS for Tests

Source         DF      Seq SS     Adj SS     Adj MS       F      P
Intake          1     1215.24    1215.24    1215.24   18.90  0.001
Cage(Intake)    4      377.42     377.42      94.35    1.47  0.272
Error          12      771.68     771.68      64.31
Total          17     2364.34
```

The output confirms the test used, the names of the factors, how many levels there are for each factor and the integer labels used. Check these are correct before looking at the ANOVA table. The table itself is very similar to the 'ideal' table given in the example. The degrees of freedom may seem a little odd in a nested analysis. The main factor will have one fewer degrees of freedom than the number of factor levels, as usual. The nested factor will have one less than the number of groups *for each level of the main factor*. In the example there are three cages so 2 degrees of freedom for each of two levels of the main factor will give 4 degrees of freedom. The important column of the table is the final one with the *P*-values (in the example there is no significant effect of 'Cage(Intake)', i.e. cage nested within intake, but a highly significant result for 'Intake').

Excel There is no easy way to carry out this type of nested analysis in Excel.

The tests 2: tests to look at relationships

Is there a correlation or association between two variables?

Observations assigned to categories

If observations are given a qualitative value then the data are said to be categorical (i.e. they have been assigned to categories). There are many occasions when it is not possible to express data in any other way (such as when scoring flower colours or species). There will be other circumstances when the categories are for convenience only, and are achieved by imposing a set of categories on a continuous scale. This can either be an arbitrary scale (e.g. the effect of an illness from 'well' through 'showing symptoms' to 'dead') or a scale that could be measured (e.g. dividing tree heights into 'short', 'medium' and 'tall').

However if the observations are assigned to categories in this way there are several tests that can be applied to determine whether the division of the observations into classes is independent or not. In other words 'are the observations for two categorical variables associated?'. The **chi-square test of association**, **Cramér coefficient** and **Phi coefficient** are considered here.

Chi-square test of association

This is one of the most widely used statistical tests of all. It is beguilingly simple and has few underlying assumptions. In fact, the test is so simple to carry out it is often not properly supported in statistical packages leaving the user to do a lot of the work. If observations can be assigned to one of two or more categories in two variables then chi-square is appropriate. The null hypothesis is that the categories in the two variables are independent (i.e. the category an observation is assigned to for one variable has no effect on the category it is assigned to for the second variable). For example if 'eye colour' and 'sex' are the two variables and individuals are assigned to either 'blue' or 'brown' and to either 'male' or 'female' then the null hypothesis is that there is no association between sex and eye colour. As there are no assumptions made about the form of the data it is a non-parametric test although it is rarely described as such (probably because there is no parametric equivalent).

(*Note*: it is important to realize that if a continuous variable is forced into a small number of categories then information is lost when the test is calculated and another measure of association is probably more appropriate.)

The chi-square test works by adding up the squared differences between the expected number of observations in a category combination and the actual observed number. The result is then looked up on a chi-square table with a number of degrees of freedom equal to one less than the number of rows multiplied by one less than the number of columns.

Expected values are calculated very simply by putting the data into a table, then totalling the observations in each row and column. The expected value for each category combination is the row total multiplied by the column total divided by the total number of observations.

There are complications. If there are any expected values lower than one the test should not be used (categories should be combined to avoid this problem). No more than 20% of the expected values should be less than five.

Chi-square tests should never be carried out on percentages or data transformed in any way. It must be carried out on frequencies (numbers of observations). When reporting the results in text it is usual to give the statistic, the degrees of freedom and some measure of the *P*-value (e.g. $\chi^2 = 15.62$, d.f. = 4, $P < 0.01$).

Example. A group of students interested in aquatic invertebrates want to determine, in the shortest possible time, whether stream velocity is related to plant growth and substrate. A qualitative survey of 50, randomly assigned, stream sections in the study area was carried out. Stream velocity was scored as 'slow' or 'fast' (if more time were available then an accurate measure of the stream velocity could be made, although it might not be very useful as such a measure is very dependent on weather conditions). The stream bed was assigned to one of four categories: 'weed choked', 'some weeds', 'shingle' and 'silt' (this is one of many possible classification systems that could be used).

The data were collected and compiled into a table as follows; numbers indicate the number of sites with each combination of categories:

		Stream-bed category			
		Choked	Weeds	Shingle	Silt
Velocity category	Slow	10	8	2	7
	Fast	2	6	10	5

The null hypothesis H_0 is that the proportion of streams in each of the 'stream bed' categories is the same for the two velocity categories. The alternative hypothesis H_1 is that the proportions vary between the two categories, i.e. there is some association with particular stream-bed categories and velocity categories appearing together more often than would be expected by chance and others appearing less often. The test will not determine which category combinations are more or less common than expected although this can be investigated by inspection of the table of raw data and expected values.

The data table needs to be processed to obtain the expected values. First the row and column totals are required.

	Choked	Weeds	Shingle	Silt	Totals
Slow	10	8	2	7	27
Fast	2	6	10	5	23
Totals	12	14	12	12	50

Then, the expected value for each of the cells can be calculated. For 'Choked' and 'Slow' the expected value is the row total (27) multiplied by the column total (12) divided by the total number of observations (50). Of course this sort of manipulation is very easy to do in a spreadsheet such as Excel.

The expected values are:

	Choked	Weeds	Shingle	Silt
Slow	6.48	7.56	6.48	6.48
Fast	5.52	6.44	5.52	5.52

In this example, the value of the chi-square test is 11.036, there are 3 degrees of freedom and the P-value is just over 0.01. This might be reported in the text of a results section as 'a chi-square test showed there was an association between stream velocity category and stream bed category ($\chi^2 = 11.036$, d.f. $= 3$, $P < 0.05$)'.

SPSS The chi-square test of association is not easy to achieve from a table of frequencies. However, if the data are arranged so that each individual is represented by a row with entries in two columns for the two factors then it is very easy to carry out.

Arrange the data in two columns with one column for each of the factors. Input the categories for each observation on a separate row with each category coded with integers starting at 1. The columns should be labelled appropriately.

The example data will have 50 rows and two columns set out using the following style:

	Velocity	Bed_cat
1	1	4
2	2	2
3	2	3
:	:	:
50	1	2

The integer codes can be used as labels for the category names. This can be done by clicking on the 'Variable View' and then clicking on the ellipsis ('...') in the 'Values' column. In the 'Value Labels' box that appears labels can be added for each category in turn. After assigning labels for all the categories click 'OK' and repeat for the other variable. Doing this before running the chi-square test will make the output much easier to interpret.

To run the test go to the 'Analyze' menu, choose 'Descriptive statistics' and then 'Crosstabs...'. Move the column containing the 'row' information into the 'Row(s):' box (in the example this was 'velocity') and the other column into the 'Column(s):' box. Then click on the 'Statistics...' button. In the options box that appears select 'Chi-square' and click 'Continue'. Then click on the 'Cells...' button. In the options box select both 'Observed' and 'Expected' in the 'Counts' area. Click 'Continue' then 'OK' in the 'Crosstabs' box. Selecting the 'Display clustered bar charts' produces a useful visual summary of the frequencies in each of the categories, but the default colours are quite awful.

The following output will appear in the 'Output' window:

Crosstabs

Case Processing Summary

	Cases					
	Valid		Missing		Total	
	N	Percent	N	Percent	N	Percent
VELOCITY * BED_CAT	50	100.0%	0	.0%	50	100.0%

VELOCITY * BED_CAT Crosstabulation

			BED_CAT				
			choked	weeds	shingle	silt	Total
VELOCITY	slow	Count	10	8	2	7	27
		Expected Count	6.5	7.6	6.5	6.5	27.0
	fast	Count	2	6	10	5	23
		Expected Count	5.5	6.4	5.5	5.5	23.0
Total		Count	12	14	12	12	50
		Expected Count	12.0	14.0	12.0	12.0	50.0

Chi-Square Tests

	Value	df	Asymp. Sig. (2-sided)
Pearson Chi-Square	11.036[a]	3	.012
Likelihood Ratio	11.945	3	.008
Linear-by-Linear Association	3.160	1	.075
N of Valid Cases	50		

a. 0 cells (.0%) have expected count less than 5. The minimum expected count is 5.52.

The first table just confirms how many rows of data have been processed by the test. The second table gives the data laid out as a contingency table. If labelling of the values for each variable has not been carried out there will only be numbers for the rows and columns. The table gives the number observed above the number expected for each category combination. Row and column totals are also given. Then comes a small table of the results of the test itself. The important line is that labelled 'Pearson Chi-Square'. The number in the 'Value' column is the value of the X^2 approximation to χ^2 (11.036 in the example). Then the degrees of freedom are given (labelled 'd.f.'). There should be 'row − 1 × columns − 1' degrees of freedom. The P-value (labelled 'Asymp. Sig. (2-sided)') appears in the last column. Ignore the other lines in this table. In the example the P-value is less than 0.05 so the null hypothesis is rejected. Inspection of the table shows that the biggest difference between the observed and expected values are in the 'Shingle' column suggesting that there is an association between shingle stream beds and fast flowing water.

The minimum expected frequency, given as 5.52 in the example, is an important number to look at as if it is less than 1 then the chi-square test should not be carried out. Caution should also be applied if more than 20% of the cells have expected frequencies of less than 5. As shown below, SPSS gives a warning if there is a problem (although it still reports the results of the test and gives a P-value):

N of Valid Cases	33		

a. 6 cells (75.0%) have expected count less than 5. The minimum expected count is 2.73.

MINITAB The chi-square test is easy to carry out from a table of counts of observations. First input the table exactly as it is laid out in the example. The columns can be labelled with the names of the categories but there is no way to label the rows. Then from the 'Stat' menu, select 'Tables' and then 'Chi-Square Test...'. In the dialogue box highlight the relevant columns and click on 'Select'. This moves the columns into the 'Columns containing the table:' box. Click 'OK'.

The example data with column labels gives the following output in the session window:

```
Chi-Square Test: Choked, Weeds, Shingle, Silt

Expected counts are printed below observed counts

         Choked    Weeds  Shingle    Silt    Total
   1        10        8        2        7       27
          6.48     7.56     6.48     6.48

   2         2        6       10        5       23
          5.52     6.44     5.52     5.52

Total       12       14       12       12       50

Chi-Sq =  1.912 +   0.026 +   3.097 +   0.042 +
          2.245 +   0.030 +   3.636 +   0.049 = 11.036
DF = 3, P-Value = 0.012
```

The output confirms the test and then gives a repeat of the table of frequencies but with the addition of row and column totals as well as the expected values. The section 'Chi-Sq' gives the value for: '(observed – expected)^2/expected' for each cell in the table and then gives the total of these values which is the statistic X^2 (estimating χ^2). The degrees of freedom are given and then the P-value. In the example the P-value is less than 0.05 so the null hypothesis is rejected. There is some association between stream-bed category and velocity category. Inspection of the 'Chi-Sq' values shows that the biggest contributors are in column 3 ('Shingle') and show that a shingle bed is far more common in a fast flowing stream (row 2) and far less common in a slow flowing one (row 1) than expected if there were no association between the factors.

Excel The spreadsheet capabilities of Excel make the chi-square test quite easy to carry out.

1 Input the frequency data (number of observations in each category combination) as a table with row and column labels exactly as in the example above.

2 Add labels for an extra row and column of totals. In the relevant cells type '= SUM' and the cell numbers required (in the example the row totals column is F so the command in cell F2 is '= SUM(b2 : e2)' and the first of the column totals is in B4 with the command '= SUM(b2 : b3)'). Once the first cell has been calculated for row and column it can be pasted to calculate the other totals. An easy way of doing this is to highlight the cell, then click and hold the small black square in the bottom right corner, and then drag across to highlight the cells to paste into. Make sure the total of row totals is calculated as well as this is the total number of observations (50 in the example).

3 A second table, the same size as the first should be labelled elsewhere on the same sheet. This is for the expected values. Once the row and column labels have been created the expected values for each cell can be calculated as '= row total * column total / number of observations' inputting the relevant cell numbers as required. In the example the top right cell expected value is calculated as '= b4*f2/f4'.

(It is useful to use the '$' as an anchor when calculating the expected values. Any cell code containing a '$' will *not* alter when the cell contents are pasted elsewhere on the spreadsheet. So when inputting the first cell calculation use '= b$4*$f2/f4' and then paste the cell across the whole table. Paste the cell across four columns and then, with four cells highlighted, paste down a row. The row containing the column totals and the column containing the row totals, as well as the cell containing the total number of observations (F4), all remain fixed.)

This should be on the spreadsheet:

Observed	Choked	Weeds	Shingle	Silt	Totals
Slow	10	8	2	7	27
Fast	2	6	10	5	23
Totals	12	14	12	12	50
Expected	Choked	Weeds	Shingle	Silt	
Slow	6.48	7.56	6.48	6.48	
Fast	5.52	6.44	5.52	5.52	

None of the expected values are below 5 so the test can proceed. If any cell had a value of less than 1 or more than 20% had values less than 5 then categories should be combined before the test is carried out.

4 Once the observed and expected tables are complete, select an empty cell anywhere on the spreadsheet where the two tables are still on the screen. Then go to 'Paste function' (f_x button), select 'Statistical' and then 'CHITEST' from the list. Click 'OK'. Then put the cell codes for the observed data (not labels) into the 'Actual_range' box. This can be done either by typing in the cell codes for top left and bottom right directly with a colon (:) between them or by selecting the 'Actual_range' box and then clicking and dragging over the cells. Do the same for the 'Expected_range' box, selecting the expected values. Then click 'OK'. The *P*-value for the test appears. In the example the *P*-value is 0.0115 which is lower than the critical 0.05 value so the null hypothesis is rejected: there is an association of stream-bed category with velocity category. Comparison of the observed and expected values indicates that 'Slow' and 'Choked' are more common than would be expected by chance, as are 'Fast' and 'Shingle'.

5 It is usual to present the results of a chi-square test giving the degrees of freedom and the value of χ^2. The degrees of freedom for the test are one less than the number of rows multiplied by one less than the number of columns (in the example $(4 - 1) * (2 - 1) = 3$). Excel provides a simple method for determining the chi-square value from a known *P*-value and degrees of freedom. Select an empty cell. Click on the 'Paste function', select 'Statistical' and then 'CHIINV'. Click on 'OK'. Input the *P*-value into the 'probability' box (or click on the cell used in step 4 above) and input the degrees of freedom, then click 'OK'. The chi-square value appears in the cell (11.03 in this example).

Cramér coefficient of association

This is a test carried out on tables of frequencies in conjunction with a chi-square test that provides additional information about the strength of the association. The statistic X^2 is used to determine significance while the Cramér coefficient (C), a measure from 0 (no association) to 1 (perfect association), is independent of the sample size. If this statistic is available as an option in the package you are using then it will allow direct comparison of the degree of association between tables.

SPSS Follow the instructions as for the chi-square test of association. In the 'Statistics' dialogue box within 'Crosstabs' select 'Phi and Cramér's V' in addition to 'Chi-square'.

MINITAB There is no direct method.

Excel There is no direct method.

Phi coefficient of association

A special case of the Cramér coefficient for 2×2 tables (i.e. there are only two categories for each of the two variables). It is used in combination with the chi-square test as it generates a value (r_ϕ) that ranges from 0 to 1, indicating a range from no association through to perfect association. It can be calculated directly after a chi-square test has been carried out as it is equal to the square root of the value of X^2 after it has been divided by the number of observations.

SPSS Follow the instructions as for the chi-square test of association. In the 'Statistics' dialogue box within 'Crosstabs' select 'Phi and Cramér's V' in addition to 'Chi-square'.

MINITAB There is no direct method. But it is a fairly easy calculation to achieve on a calculator, in Excel or in MINITAB. Note down the value of X^2 and the number of observations in total. Enable the command prompt by selecting the 'session' window, then from the 'Editor' menu select 'Enable commands'. At the MINITAB prompt type 'LET K1 = value1', where 'value1' is the X^2 from the test. Then 'LET K2 = value2', where 'value2' is the number of observations. To calculate the phi coefficient, type 'LET K3 = SQRT(K1/K2)', and to display the result, type 'Print K3'.

Excel If the value of X^2 is in cell E1 and the number of observations is in C3, then the formula: '= SQRT(E1/C3)' will give the phi coefficient.

Alternatively, if the 2×2 contingency table is laid out in cells A1, A2, B1 and B2 then the phi coefficient can be calculated directly by the rather horrible looking formula:

'= ((A1 * B2) − (A2 * B1))/(SQRT((A1 + B1) * (A2 + B2) * (A1 + A2) * (B1 + B2)))'.

Observations assigned a value

If each individual is assigned a meaningful numerical value for two variables there are several tests that may be applied to determine whether the two sets of observations are associated or correlated, the strength of the correlation and whether it is significant or not. Four tests are considered here, the **Pearson product–moment correlation**, the **Spearman rank-order correlation**, the **Kendall rank-order correlation** and **regression**. The most appropriate test is determined by the distribution and quality of the data.

'Standard' correlation (Pearson product–moment correlation)

When a correlation is mentioned it is almost invariably the Pearson product–moment correlation that is in mind. The statistic, r, to estimate the true correlation, ρ (rho), produced by the test ranges from –1, through 0, to 1. and describes a range of associations between two variables from 'perfect negative correlation', through 'no correlation' to 'perfect positive correlation'. This test is very widely applied, perhaps too widely as it has some rather severe assumptions about the distribution of the two variables being investigated. Both variables must be measured on a continuous scale and both must be normally distributed (often termed a bivariate normal distribution). If these assumptions do not apply the Spearman rank-order correlation should be used instead.

When quoting the results of this test in a report it is usual to use a scatter-graph (without trend line) and the form: 'Pearson product–moment correlation indicates a significant positive association between x and y ($r = 0.51$, d.f. = 22, $P < 0.05$)'.

Two words of caution:
1 It is quite rare to find two variables that are normally distributed and therefore suitable for a Pearson correlation. Test the data to see if it follows a normal distribution. Consider the alternatives.
2 The statistical significance of correlation is not a good guide to the *real* significance of the correlation. With large sample sizes the value of r required to achieve statistical significance (i.e. to show that there is some relationship between the two variables) is rather low. It is perhaps better to use the value of r^2 as an indicator of the real significance as this value shows the amount of variation in one variable explained by the other.

Example. A marine biologist working on Adélie penguins (*Pygoscelis adeliae*) has measured the sizes of birds forming pairs. The measure used is the length of a bone in the leg which is known, from previous studies, to be a good indication of size. It is measured to the nearest 0.1 mm. The null hypothesis is that male size is not correlated with female size. Unfortunately data could only be collected from six pairs.

Pair	Female	Male
1	17.1	16.5
2	18.5	17.4
3	19.7	17.3
4	16.2	16.8
5	21.3	19.5
6	19.6	18.3

It is assumed that both variables are normally distributed. This data set is a little small to prove this although a larger sample of the population might be (see 'Do frequency distributions differ?' in the previous chapter, p. 66). In this case the null hypothesis is rejected as there is a significant positive correlation between male and female size indicating that there is positive, assortative mating in this species. The value of r is 0.88 and r^2 is 0.77 indicating that 77% of the variation in the size of one sex is explained by the size of the other.

SPSS This is one of the easiest tests to carry out in SPSS. Input the data in two columns and add appropriate column labels. The cases (pairs in the example) do not require a separate column. From the 'Analyze' menu, select 'Correlate' then 'Bivariate…'. In the dialogue box move the names of the two variables into the 'Variables:' box and make sure that the 'Pearson' option and 'Two-tailed' are checked. Click 'OK'.

The following output appears:

Correlations

Correlations

		FEMALE	MALE
FEMALE	Pearson Correlation	1.000	.881*
	Sig. (2-tailed)	.	.020
	N	6	6
MALE	Pearson Correlation	.881*	1.000
	Sig. (2-tailed)	.020	.
	N	6	6

*. Correlation is significant at the 0.05 level (2-tailed).

The output gives a matrix of correlation coefficients, degrees of freedom and P-values, although only one set is actually useful. The correlations of 'x' with 'x' and 'y' with 'y' (e.g. female with female in the example) unsurprisingly report a perfect correlation of '1.000'. The important section is the correlation of 'x' with 'y' (female with male) in the example. The statistic is given first ($r = 0.881$ in the

Fig. 8.1 Bivariate correlation in MINITAB. Each pair of variables should be highlighted and selected to move them into the 'Variables:' box.

example), then the *P*-value (labelled 'Sig. (2-tailed)' and 0.020 in the example) and finally the numbers of pairs of observations. As *r* is positive it shows that there is positive size assortment of individuals in pairs in the example (i.e. large females tend to be paired with large males) and the association is strong enough to reject the null hypothesis and accept the alternative hypothesis.

MINITAB Input the data in two columns using one row for each pair of observations. Label the columns appropriately. From the 'Stat' menu select 'Basic statistics' then 'Correlation...'. Highlight the names of the two variables then press the select button (Fig. 8.1).

Click 'OK'.

(*Or type* 'corr c1 c2' *at the MTB > prompt in the session window.*)

The following output appears:

Correlations: Female, Male

```
Pearson correlation of Female and Male = 0.881
P-Value = 0.020
```

The test reports the value of r and the P-value. In the example this indicates that there is a strong positive association between the two variables — indicating that there is positive assortative mating. In the example the value of P is well below 0.05 indicating that this is a significant association, even with a very small sample size and the null hypothesis of no association should be rejected. However, as noted above, it is important to be cautious about the significance of correlations, especially with large samples.

Excel There are two methods, the first requires the 'Data analysis toolpak' to be installed, the second does not.

Method 1. Arrange the data in two columns with appropriate column labels. From the 'Tools' menu select 'Data analysis...' and then 'Correlation'. Click 'OK'. The 'Input range' box should contain the cell code for the top left and bottom right of the area containing the data separated by a colon (:). This can be done most easily by selecting the 'Input range' box, then clicking on the top left cell and dragging to the bottom right. Data in the example are arranged in columns, and that option should be selected by default. If the variable labels are included in the selected cells, the 'Labels in first row' box should be checked. The output can either appear on a separate sheet or elsewhere on the same sheet. If on the same sheet the 'Output range' option needs to be selected and the cell indicating the point at the top left of the output put into the box. Click 'OK'.

This table appears:

	Female	Male
Female	1	
Male	0.88132	1

Method 2. Arrange the data in two columns (or two rows). Select an empty cell then click on 'Paste function', then 'Statistical', and select 'Correl'. Click 'OK'. Input the range of cells containing the data for the first variable into the 'array1' box (this can be done by either typing in the first and last cell with ':' between them or by clicking on the first cell and dragging the pointer to the last cell). Put the range for the second variable in the 'array2' box and click 'OK'. The r-value appears in the cell.

Either method will produce the same number although the output from method 1 is more extensive. The value of the statistic r is given. It is 0.88132 using the example data, indicating a strong positive association of the two sets of observations. The r^2-value can be determined by selecting an empty cell and typing '=0.88132 ^ 2' (the '^' symbol is often used to mean 'to the power of'). If the P-value associated with this value of r is required this can be determined using regression (regress one variable on the other and inspect the P-value reported in the ANOVA table section of the output; see the regression section later in this chapter for further details).

Spearman rank-order correlation

This is one of two commonly used non-parametric equivalents of the Pearson product–moment correlation. The statistic it gives is called r_s and ranges from -1, through 0, to 1, indicating 'perfect negative correlation', 'no correlation' and 'perfect positive correlation'. Although this is superficially the same scale as the Pearson r it is not advisable to compare the results from this test directly with values of r.

As long as there are two observations for each individual and the observations are measured on a scale that can be put into a meaningful rank order this test is appropriate. The Spearman correlation is much more conservative than the Pearson.

Example. The same penguin example will be used as for the Pearson product–moment correlation used earlier. It is quite possible that a researcher will use the Spearman rank-order correlation rather than the Pearson if there is a doubt that data are appropriate for the Pearson test.

SPSS Arrange the data in two columns with each row representing an individual (in the example the 'individuals' being considered are actually pairs of birds). Label the columns appropriately. From the 'Analyze' menu select 'Correlate', then 'Bivariate...'. Highlight the names of the two variables and move them into the 'Variables:' box. Make sure that 'Spearman' is checked (Pearson is checked by default) and that the 'Test of Significance' is 'Two-tailed'. Click 'OK'.

The following output appears:

Nonparametric Correlations

Correlations

			FEMALE	MALE
Spearman's rho	FEMALE	Correlation Coefficient	1.000	.771
		Sig. (2-tailed)	.	.072
		N	6	6
	MALE	Correlation Coefficient	.771	1.000
		Sig. (2-tailed)	.072	.
		N	6	6

This confirms the test used and then gives the value of r_s, the number of pairs of observations and the P-value associated with the r_s value (labelled 'Sig. (2-tailed)'). In the example although the statistic is clearly showing a large positive association it is not large enough to be deemed significant because the sample size is rather small. However the P-value is sufficiently close to 0.05 to warrant further investigation.

MINITAB This test cannot be carried out directly. The data first have to be put into rank order and then a Pearson correlation carried out on the ranked data.

1 Input the data in two columns. Label the columns appropriately. If there are any individuals where one or both of the observations are missing these should be omitted from the data set. From the 'Manip' menu select 'Rank...'. Put the name of the first variable into the 'Rank data in:' box and put a new name (e.g. 'RankF') in the 'Store ranks in:' box. Click 'OK'. A new column will appear with the original data replaced with integers starting at '1' for the smallest value. Repeat for the second variable.

2 Carry out a normal Pearson correlation on the *ranked* data. From the 'Stat' menu select 'Basic statistics', then 'Correlation...'. Highlight the two *ranked* variables in the list on the left and click on select to move them into the 'Variables:' box. Click 'OK'.

This output appears:

Correlations: RankM, RankF

```
Pearson correlation of RankM and RankF = 0.771
P-Value = 0.072
```

The test indicates that a Pearson correlation has been carried out (the package does not 'know' that ranked data are being used). The correlation reported is the value of r_s. In this case it shows that there is a strong positive association between male and female size. But the significance of 0.072 is not small enough for us to reject the null hypothesis.

Excel There is no direct way of carrying out this test even with the 'Data analysis toolpak' installed. However it is possible to carry out the test by first ranking the data and then performing a normal correlation.

1 Input the data in two columns with a pair of observations on each row. Do not include any rows where one or both of the observations are missing. Use the cells at the top of the columns for appropriate labels. To rank the data first select a cell in an empty column in the same row as the first row of data. To use the 'rank' operation it is probably easiest to simply type '= RANK(A2,A\$2 : A\$7)' rather than use 'Paste function'. This command asks that the cell is given the rank of cell A2 in the range of cells A2 to A7 (the six cells containing the female sizes in the example). The inclusion of the '\$' symbol makes it very easy to use the copy and paste commands to rank the whole data set as it anchors the row numbers, thus preserving the correct range during the pasting. First select the cell, then click on the small black square in the bottom right and drag to the cell in the next column to the right level with the *last* observation in the data set. This will place the appropriate 'RANK' syntax in each cell without having to type in anything else (in the example it left '= RANK(B7,B\$2 : B\$7)' in the bottom right cell). Label the columns containing the ranks appropriately.

Fig. 8.2 Spearman rank correlation in Excel. The data have to be ranked first and then a normal, Pearson product–moment correlation is carried out on the ranked data.

2 Carry out a Pearson correlation on the *ranked* data. Select an empty cell. Click on 'Paste function' and select 'CORREL' from the 'Statistical' section of the list. Click 'OK'. Input the range of cells containing the data for the first variable into the 'Array1' box (this can be done by clicking on the first cell and dragging the pointer to the last cell). Put the range for the second variable in the 'Array2' box and click 'OK' (Fig. 8.2). The r_s value appears in the cell.

In the example the value of r_s is 0.7714, indicating a strong positive association of ranked female and ranked male size. No value of significance is given.

Kendall rank-order correlation

This is a second, and slightly less widely used, non-parametric correlation. It is sometimes called Kendall's tau and the statistic produced is usually denoted as T. As with the Spearman correlation, the test can be carried out on any data set where there are two observations for each individual and the data can be put into a meaningful rank order. Like other measures of association, T ranges from -1, through 0, to 1, indicating 'perfect negative correlation', 'no correlation' and 'perfect positive correlation'. Although this is the same as the Pearson r and the Spearman r_s it is not advisable to compare the results from this test directly with either.

The only slight advantage of the Kendall correlation over Spearman is that T can be used in partial correlation, which r_s cannot.

It is advisable to use caution when interpreting the significance of a value of *T* as when sample sizes are very large there is likely to be a *P*-value of less than 0.05 even though the association is only slight.

Example. The same example as used for the Pearson and Spearman tests.

SPSS As for the Spearman correlation, arrange the data in two columns with a row for each 'individual'. Label the columns appropriately. From the 'Analyze' menu select 'Correlate' then 'Bivariate...'. Select the two variables and move them into the 'Variables:' box. Select the 'Kendall's tau-b' option from the 'Correlation coefficients:' list. Make sure 'Test of Significance' highlights 'Two-tailed'. Click 'OK' to run the test.

The following output appears:

Nonparametric Correlations

Correlations

			FEMALE	MALE
Kendall's tau_b	FEMALE	Correlation Coefficient	1.000	.600
		Sig. (2-tailed)	.	.091
		N	6	6
	MALE	Correlation Coefficient	.600	1.000
		Sig. (2-tailed)	.091	.
		N	6	6

This confirms the test used and then gives the value of *T*, the number of pairs of observations and the *P*-value associated with the *T*-value (labelled 'Sig. (2-tailed)'). In the example, although the statistic is clearly showing a large positive association, it is not large enough to be deemed significant because the sample size is rather small. However the *P*-value is close enough to 0.05 to be worthy of further investigation.

Note that the value of *T* is rather less than the value of r_s in the Spearman test although both used the same data. This highlights the fact that the two statistics should not be compared directly.

MINITAB A Kendall rank correlation is not possible in MINITAB; use the Spearman rank correlation instead.

Excel There is no direct way of carrying out this test. Use the Spearman rank-order correlation instead.

Regression

The use of regression usually implies that a prediction of one value is being attempted from another, e.g. cause and effect. Regression analysis is considered in the next section. However, as most regression output gives a P-value and an r^2 value, the similarity to the Pearson product–moment correlation is very great. The P-value given in a standard linear regression is the probability that the best-fit slope of the relationship between two variables is actually zero. In a comparison with the Pearson statistic, this translates to the probability that there is no relationship (i.e. $r = 0$). Regression analysis usually considers a second null hypothesis: 'the value of y is zero when x is zero'. This translates to a test of whether the best-fit line through the data set passes through the origin. It is often labelled as a test of the intercept.

The advantage of using regression rather than a Pearson correlation is that the assumption that both variables are distributed normally is lifted. The assumptions are different, although slightly less restrictive. For example regression assumes that the 'x' ('cause') values should be measured without error; that the variation in the 'y' ('effect') is the same for any value of 'x'; that the 'y' values should be normally distributed at any value of 'x'; and, for linear regression, that the relationship between the two variables can be described by a straight line. Of these the assumption that variance in 'y' is the same for all values of 'x' is probably the least likely to be true. It is usual for variance in 'y' to increase as the 'x' ('cause') variable increases.

If you decide to use regression to determine the association between two variables please use great caution because the implication is that one of the variables in some way depends on the other. Also one of the underlying assumptions of regression is that the values of the 'cause' variable are in some way set, or chosen, by the investigator — clearly this is not the case if the observations are taken at random.

Example. Again we consider the penguin pairs used as the example throughout this section. The researchers were testing the null hypothesis that male and female birds were forming pairs independent of their size. The alternative hypothesis was that there was an association (either positive or negative) of male and female sizes in pairs. Framed in this way the hypothesis is not suitable for regression. But, if the penguins form pairs by a choice of one sex for another then it might become more like a regression problem. If females actively chose males then the null hypothesis can be framed in regression terms as 'male size does not depend on female size'. However, even if male size can be said to *depend* on female size there is still the problem that the female sizes were not set or chosen by the investigator.

Is there a 'cause and effect' relationship between two variables?

There are many circumstances where it is clear that one set of observations in some way depends on another. In this section there will be two observations for each 'individual' with one observation being the 'cause', 'x', 'predictor' or 'independent' variable that is set, or chosen, by the experimenter and the other being the 'effect', 'y' or 'dependent' variable which is never set by the experimenter. There are a variety of methods that can be applied to determine the form and strength of the relationship between the 'cause' and 'effect' that make different assumptions about the variables and the form of the relationship between them. Five tests are considered here: **linear regression, Kendall robust line-fit method, logistic regression, model II regression** and **polynomial regression**.

'Standard' linear regression (a.k.a. model I linear regression)

This is a very widely used statistic in biology. It is also, possibly, the most abused statistic in biology because the assumptions of the test are often flouted. Linear regression is an extremely powerful and useful technique that determines the form and strength of a relationship between two variables. It is used if the intent is to be able to predict a value for y ('effect' or 'dependent') from a given value of x ('cause', 'predictor' or 'independent'). There are several components to any output. The 'slope' is the slope of a straight line of best fit drawn through the set of points with coordinates defined by the two variables. Slope can be positive or negative indicating an increase or decrease of y with increasing values of x. The slope can, theoretically, take any value. A slope of zero indicates no change in y with x. The second component of the output is the 'intercept' or 'constant'. This is the predicted value of y when x is equal to zero. The slope is often called 'b' or 'm' and the intercept 'a' or 'c' or 'constant'. Both the slope and intercept are usually quoted with some measure of their variability (e.g. a 95% confidence interval or standard deviation). There is often a significance test result given with regression output. This is a test of whether the slope is zero or not (i.e. testing a null hypothesis that $b = 0$). If the P-value is less than 0.05 this should be interpreted as an indication that the slope is significantly different from zero, indicating that there is a relationship between the x and y variables.

Linear regression makes many assumptions about the data sets. Important assumptions include: that the values of x are measured without error; that the values of x are chosen or set by the experimenter; that the relationship between x and y is best fitted by a straight line ($y = a + bx$); that the variation in y is the same for all values of x; and that y is normally distributed for any value of x.

Tip: if you are unsure which of the two variables is the x ('cause') variable then linear regression is almost certainly not appropriate.

Prediction. Once a best-fit line has been determined then a value for the 'effect' can be predicted for any value of the 'cause'. In practice it is unwise to use values of the 'cause' variable that are beyond the range of the data used to fit the line as the shape of the relationship is unlikely to be the same across all values of the 'cause'. Although it may appear tempting, you should *never* use an observation of the 'effect' to make a prediction of the 'cause' by simple rearrangement of the algebra of the best-fit line.

Interpreting r^2. One commonly used output from linear regression is the value of r^2. This is often expressed as a percentage and is described as the amount of variation explained by the regression (i.e. how much of the variation in the 'effect' can be accounted for by using the relationship between 'cause' and 'effect'). Another way to interpret r^2 is to consider that a prediction of the 'effect' can be made from its overall mean, and the percentage improvement in prediction in the 'effect' variable that results from the regression is given by r^2.

Comparison of regression and correlation. The assumptions of linear regression are very different to those of correlation. 'Standard' correlation assumes that both x and y are normally distributed. For this reason it is tempting to use regression in lieu of correlation when variables are not normally distributed. Do not do this — try an alternative correlation instead because regression makes different assumptions that are unlikely to be true if correlation was the preferred statistical technique.

Residuals. The variation in y not accounted for by the best-fit line relationship between x and y is called the residual variation. For each observation of x there is a predicted value on the line of y. Because y varies, any point is unlikely to lie exactly on the fitted line. The *vertical* distance from the point to the line is the *residual* for that point. It is often helpful to examine the residuals by plotting them against x. This is offered as an option in most statistical packages. If the relationship of x and y is really a straight line, or there is no relationship at all, then the residuals will be scattered without pattern for all values of x. However, if the relationship is really a curve then the residuals of a best-fit straight line will show this to be the case — most of the residuals will be negative (or positive) at the ends of the line and positive (or negative) in the middle. If this is the case, then either a polynomial regression might be a better option than a linear regression or the data should be transformed.

Also, one of the assumptions of regression is that there is the same variation in y for all values of x. If the residuals are all small at one end of the range of x values and large at the other this would indicate that this assumption has been violated.

Confidence intervals. The confidence intervals (CI) attached to slope and intercept allow a range of possible lines to be drawn with boundaries enclosing the range within which 95% (or 90%, if preferred) of the lines of best fit will appear. As this range of possible lines is, in some way, anchored to the y-axis at the point where x is zero and comprises a range of lines of different slope, the 95%

confidence limits of the lines are *not* straight lines. They are always curved lines that are closer to the best-fit line in the middle of the range of values of x and further away from it at the extremes. This indicates that you can be more confident about the predictions of the regression in the middle of the data.

Prediction interval. This is a range within which 95% of the values of y are predicted to occur for any given value of x. The 95% prediction intervals (PI) will always be further away from the best-fit line than the 95% confidence interval for the line. They do not run parallel to the best-fit line but, like the confidence interval for the line, are narrower for mid-range values of x.

Example. A team of researchers is investigating the uptake of an experimental drug through the stomach. They suspect that the acidity of the stomach will affect uptake and propose an experiment to determine the uptake across a range of acidities. Using preparations of sheep stomach and a fixed concentration of the drug in solution, a range of pH is prepared and the passage of the drug through the stomach monitored. This is suitable for regression as the acidities ('cause') are set by the experimenters and we can assume that they are measured without error.

	pH					
	0.6	0.8	1.0	1.2	1.4	1.6
Uptakes	11.32	11.29	11.37	11.32	11.32	11.49
	11.31	11.22	11.40	11.31	11.36	11.52
	11.22	11.18	11.38	11.35	11.40	11.38
	11.23	11.21	11.37	11.32	11.35	11.49

It is usually best to plot the data as a scattergraph at this point to get a feel for the form of the relationship. This has been done in Excel (Fig. 8.3).

SPSS There are many options that can be applied to enhance the output of the standard linear regression in SPSS. First, the minimum output will be considered and then options that might be useful to help interpret the form of the relationship between the two variables.

Input the data in two columns and label them appropriately. There should be one row for each observation (two numbers). From the 'Analyze' menu select 'Regression' and 'Linear...'. Move the column containing the y or 'effect' variable into the 'Dependent:' box. Move the x or 'cause' variable (the one set or chosen) into the 'Independent(s):' box. Ignore all of the possible options, just click 'OK'.

Using the example data the following output appears:

Regression

Variables Entered/Removed[b]

Model	Variables Entered	Variables Removed	Method
1	PH[a]	.	Enter

a. All requested variables entered.

b. Dependent Variable: UPTAKE

Model Summary

Model	R	R Square	Adjusted R Square	Std. Error of the Estimate
1	.758[a]	.574	.555	5.893E-02

a. Predictors: (Constant), PH

ANOVA[b]

Model		Sum of Squares	df	Mean Square	F	Sig.
1	Regression	.103	1	.103	29.654	.000[a]
	Residual	7.641E-02	22	3.473E-03		
	Total	.179	23			

a. Predictors: (Constant), PH

b. Dependent Variable: UPTAKE

Coefficients[a]

Model		Unstandardized Coefficients		Standardized Coefficients	t	Sig.
		B	Std. Error	Beta		
1	(Constant)	11.127	.041		274.296	.000
	PH	.192	.035	.758	5.446	.000

a. Dependent Variable: UPTAKE

Even this minimum level of output is rather bewildering. First the output confirms the test used and then the names of the x variables (just one in most cases). Then comes a 'Model Summary' that gives the value of r (labelled 'R') and r^2 (labelled 'R Square'). The other columns give a more conservative estimate of r^2 and a measure of the confidence in r^2.

Next is an ANOVA table comparing the best-fit line against a null hypothesis of no relationship (e.g. slope is zero). In this table the F-ratio ('F') and its associated P-value (labelled 'Sig.') is given. In the example the P-value is given as '.000' meaning $P < 0.001$ which is highly significant. There is very little doubt that a significant amount of the variation in y is explained by x.

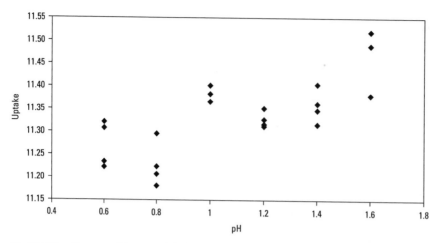

Fig. 8.3 Visualization of data suitable for regression using Excel. There were six pH levels set by the experimenter and four observations of uptake made at each level. This sort of plot gives a feel for the variation and trend in the data before any regression analysis is undertaken. It is right to plot the data like this before a trend line is added, as the line will draw the eye.

Finally come the estimates for the values of the two 'coefficients' (i.e. the slope and the intercept). The intercept is given first in a row (labelled '(Constant)' in the example) giving a value with standard error and then a t-test of the null hypothesis that the intercept is zero. The next line (labelled 'PH' here) gives a value for the slope, and is followed by an estimate of the standard error associated with the slope and then a t-test of the null hypothesis that the slope is equal to zero (this is essentially a repeat of the result of the ANOVA table).

In this example the slope of the relationship between pH and 'Uptake' is 0.192 and the intercept (value of 'Uptake' when pH is zero) is 11.127. Both these values are significantly different from zero. The best-fit line of 'uptake = $0.192 \times$ pH + 11.127' explains about 57% of the variation in the observations of 'Uptake'. The best-fit line can be, theoretically, applied to any value of x ('PH'), within the range used in the analysis, to make a prediction of y ('Uptake').

Useful options to help with the interpretation of the regression relationship include a plot of the residuals. From the 'Linear Regression' dialogue box choose 'Plots'. Then put the 'Dependent' into the 'X:' box and '*ZRESID' into the 'Y:' box. This will generate a plot of the residuals plotted against the values of the 'effect' variable. If there is a clear pattern in this plot then it is likely that one of the assumptions of the regression has been violated. For example, if the true relationship is a curve this will show up in this plot as a curved shape.

MINITAB Put the data in two columns: one for the x values and one for the y. Label the columns appropriately. From the 'Stat' menu select 'Regression', then 'Regression...'. In the dialogue box that appears move the column containing the y values into the 'Response:' box and the column of x values into the 'Predictors:' box. Ignore the array of additional options for the moment. Click 'OK'. Note that MINITAB calls the y or 'effect' variable the 'response' and the x or 'cause' variable the 'predictor'.

(Or, if the command interface is enabled, and assuming the 'y' values are in c1 and the 'x' values in c2, type 'Regress c1 1 c2' at the MTB > prompt in the session window.)

This output appears:

```
Regression Analysis: uptake versus pH

The regression equation is
uptake = 11.1 + 0.192 pH

Predictor       Coef     SE Coef        T        P
Constant     11.1270      0.0406   274.30    0.000
pH           0.19179      0.03522     5.45    0.000

S = 0.05893    R-Sq = 57.4%    R-Sq(adj) = 55.5%

Analysis of Variance

Source          DF        SS        MS        F        P
Regression       1   0.10299   0.10299    29.65    0.000
Residual Error  22   0.07641   0.00347
Total           23   0.17940
```

The output first confirms the test and then gives the best-fit line of the two variables. This describes both the intercept (11.1 in the example) and the slope (0.192 in the example). Then comes a section repeating the two parameters where the intercept is labelled as 'Constant'. A standard deviation is given for each, along with a t-test for the null hypothesis that the parameters are zero and the P-value (labelled 'P') associated with the t-test (labelled 'T'). In the example both the slope and intercept are highly significantly different from zero. Next comes a line that reports the proportion of the variance in the y values that is explained by the best-fit line. In this case the r^2 value (labelled 'R-Sq') is 57.4%.

Finally comes an analysis of variance table that compares the variation explained by the best-fit line with the residual variation. The null hypothesis is that none of the variation in y is explained by the regression line. In this case the relationship is highly significantly different from zero and P is given as '0.000' although this should be reported as $P < 0.001$.

There are many optional extras that can be added to the basic regression analysis. Some of these extras are valuable for exploring the relationship between the x and y variables. The plot of 'Residuals versus fits' can often be revealing and is a good way to detect whether the linear relationship would be better fitted to a curve.

Excel Regression is only possible if the 'Data analysis toolpak' has been installed. Put the data into two columns: one for the 'effect' and one for the 'cause'. Each row should represent a single observation (i.e. two values). Use the cells above the data to label the columns appropriately. From the 'Tools' menu select 'Data analysis...', then select 'Regression' from the long list and click 'OK'.

The 'Input Y Range:' box should contain the code for the first and last cells with the list of y values (including the label). This can be most easily done if you click in the box to move the cursor there, then click on the top of the list of data and drag to the bottom of the list. Repeat this for the 'Input X Range:' box with the x values (these are the ones set or chosen by the experimenter). If the labels have been included then make sure that the 'Labels' option is checked. Note that Excel asks for the y ('effect') values first, although it might be easier for graph drawing to have the x values to the left of the y values.

To define where the output is to appear choose either 'Output range:' and point at the top left corner of where the output will appear or leave the selection as 'New worksheet ply:' which will put the output in a fresh worksheet. Click 'OK'.

The example data produces this output:

SUMMARY
OUTPUT

Regression Statistics	
Multiple R	0.757686
R Square	0.574088
AdjustedR Square	0.554728
Standard Error	0.058933
Observations	24

ANOVA

	df	*SS*	*MS*	*F*	*Significance F*
Regression	1	0.102989	0.102989	29.65382	1.80402E-05
Residual	22	0.076407	0.003473		
Total	23	0.179396			

	Coefficients	*Standard Error*	*t Stat*	*P-value*	*Lower 95%*	*Upper 95%*	*Lower 95.0%*	*Upper 95.0%*
Intercept	11.12695	0.040566	274.2959	2.24E-40	11.04282459	11.21108	11.04282	11.21108
pH	0.191786	0.035219	5.445532	1.8E-050	.118746081	0.264825	0.118746	0.264825

This output is rather confusing. First the output confirms that regression statistics are being reported. Then come four lines giving the strength of the relationship between y and x. This is effectively the r of the Pearson correlation. The 'R Square' (usually given as 'r-square') can be interpreted as the proportion of the variation in y explained by the best-fit line. In the example 57% of the variation in y is explained by the line.

Next is an ANOVA table comparing the best-fit line against a null hypothesis of no relationship (i.e. the slope is zero). In the example the P-value (labelled as 'Significance F') is given as 1.80402E-05 meaning 0.000018 or $P < 0.0001$, which is highly significant. There is very little doubt that a significant amount of the variation in y is explained by x.

Finally comes the estimates for the two parameters in a separate table (i.e. the slope and the intercept). The first row refers to the intercept giving a value with standard error and then a t-test of the null hypothesis that the intercept is zero with the associated *P*-value (ludicrously small in the example), and then estimates of the 95% confidence intervals for the intercept (in the example the intercept is 95% likely to lie between 11.04 and 11.21). The slope is given on a row labelled 'pH' in the example. The estimated value for the slope is given, and is followed by an estimate of the standard error associated with the slope and then a *t*-test of the null hypothesis that the slope is equal to zero with the *P*-value (this is essentially a repeat of the result of the ANOVA table), followed by 95% confidence intervals for the slope.

In this example the slope of the relationship between 'pH' and 'uptake' is 0.1918 and the intercept (value of 'uptake' when 'pH' is zero) is 11.13. Both these values are significantly different from zero. The results can be interpreted as a best-fit line of 'uptake = 0.1918 × pH + 11.13' explaining 57% of the variation in the observations of 'uptake'. This best-fit line could be, theoretically, applied to any value of 'pH' to predict 'uptake'.

Options to try in Excel regression include the residual plot and table of residuals, where the residual value (distance from the best-fit line) is given for every data point. The residual plot can be used to detect systematic deviations from the assumptions of a linear regression: e.g. larger deviations at higher values of *x*, or if the true line is a curve.

Kendall robust line-fit method

This is a simple non-parametric test that can be used instead of normal regression. It is unlikely to be supported by a statistical package although it is likely to be possible to calculate using simple spreadsheet manipulations. The idea is to calculate the slope of the line between every possible pair of *x* and *y* points. This will potentially generate a large number of slopes (e.g. 10 slopes from five points, or 45 slopes from 10 points). The median slope is then selected as the best estimate of *b*.

Once the median slope has been determined the intercept can be calculated by taking the slope back from every observation to '*x* = 0'. This gives an intercept for every point. The median intercept value should be used.

This test makes very few assumptions about the data other than it is measured on a meaningful scale.

Logistic regression

This is a special form of regression analysis that is used when the 'dependent' or 'effect' variable can only be classified into groups (many packages limit this to two groups). It is a regression analysis that uses the proportions of the two possibilities of the 'dependent' converted to a logit to calculate the relationship with the 'independent' or 'cause' variable, and is particularly useful when there are proportions near 0 or 1.

Unlike linear regression, the logistic regression does not require the 'cause' ('independent' or *x*) to be on a meaningful scale. There can be as few as two points on the x-axis (e.g. male and female) and a logistic regression is still possible. Logistic regression uses logit transformations (q.v.) of the 'effect' ('dependent' or y) values, although the packages make this transformation for you.

Logistic regression is a very powerful technique and can be used as a classification tool in a similar way to discriminant function analysis. The example below is a very simple problem for logistic regression. The problem is a similar one to a linear regression but is using a 'response' or 'effect' variable that can only be in one of two categories. Logistic regression can also be used as a form of multiple regression with many 'cause' or 'predictor' variables.

Example. The null hypothesis is that the prevalence of a plant virus is unaffected by shading conditions. Researchers were able to estimate canopy cover and, by using a light meter, identified seven shading levels. Within each shade level 10 plants were selected at random and the presence or absence of the virus on each plant recorded. The results are tabulated below.

Shade category	1	2	3	4	5	6	7
Plants with virus	2	4	4	4	6	6	8
Plants without virus	8	6	6	6	4	4	2

SPSS The data should be entered as 70 rows of data with one row for each individual. There should be two columns, one for shade category and one for virus coded at either 0 or 1. From the 'Analyze' menu, select 'Regression' and then 'Binary Logistic . . .'. In the dialogue box move the 'effect' variable ('Virus' in the example) to the 'Dependent:' box. Move the 'cause' variable to the 'Covariates:' box. Click 'OK' to continue.

Masses of output is produced. Here is the most important bit:

Classification Table[a]

			Predicted		
			VIRUS		Percentage
Observed			0	1	Correct
Step 1	VIRUS	0	26	10	72.2
		1	14	20	58.8
	Overall Percentage				65.7

a. The cut value is .500

Variables in the Equation

		B	S.E.	Wald	df	Sig.	Exp(B)
Step 1[a]	SHADE	.370	.134	7.634	1	.006	1.447
	Constant	-1.544	.598	6.672	1	.010	.214

a. Variable(s) entered on step 1: SHADE.

The first table shows how many of the observations are correctly assigned to each category using the model generated by the logistic regression. In the example of the 34 plants without virus there were 26 (or 72%) correctly assigned to that category by the model. The second table is very similar to a standard regression table. The 'Shade' line is the test of the 'slope' and can be interpreted as a test of whether accounting for shade reduces the number of observations misclassified. In the example this has a *P*-value ('Sig.') of 0.006 indicating that the null hypothesis should be rejected and that shade does indeed have a very significant effect on the presence of virus. The second line is a test of the 'intercept' and is only useful if you are interested in the proportion of plants infected when the shade category is zero (i.e. not relevant in this case).

MINITAB The data can be entered as 70 rows of data with one row for each individual. There should be two columns, one for shade category and one for virus coded at either 0 or 1. Alternatively the data is entered as a frequency table. For this example I assume one row per individual.

From the 'Stat' menu, select 'Regression' and then 'Binary Logistic Regression'. Move the 'effect' variable ('Virus' in the example) to the 'Response:' box and the 'cause' variable ('Shade' in the example) to the 'Model:' box. Click 'OK'.

Here is a portion of the output that appears:

Binary Logistic Regression: virus versus shade

Logistic Regression Table

Predictor	Coef	SE Coef	Z	P	Odds Ratio	95% CI Lower	Upper
Constant	-1.5436	0.5976	-2.58	0.010			
shade	0.3697	0.1338	2.76	0.006	1.45	1.11	1.88

Log-Likelihood = -44.221
Test that all slopes are zero: G = 8.541, DF = 1, P-Value = 0.003

Table of Observed and Expected Frequencies:

Value	Group 1	2	3	4	5	6	7	Total
1								
Obs	2	4	4	4	6	6	8	34
Exp	2.4	3.1	3.9	4.8	5.8	6.6	7.4	
0								
Obs	8	6	6	6	4	4	2	36
Exp	7.6	6.9	6.1	5.2	4.2	3.4	2.6	
Total	10	10	10	10	10	10	10	70

Tau-a	0.20		
Total		1224	100.0%

The first section is very similar to a standard regression table. The 'shade' line can be interpreted as a test of whether accounting for shade reduces the number of observations misclassified. In the example this has a *P*-value of 0.006 indicating that shade does indeed have a very significant effect on the presence of virus.

The 'Constant' line is a test of the 'intercept' and is only useful if you are interested in the proportion of plants infected when the shade category is zero (i.e. not relevant in this case). The second section shows the observed and expected values generated by the logistic regression for each category of the 'cause' variable. In the example for shade category 4 the model predicted 4.8 plants to be infected and 5.2 plants not to be infected, with observed values of 4 and 6. Of the 34 plants without virus there were 26 (or 72%) correctly assigned to that category by the model.

Excel There is no direct way of carrying out this test in Excel.

Model II regression

The model II regression is actually a group of analyses that make far fewer assumptions about the data than standard model I regression. The main culprits are the assumptions in standard regression that the x values are measured without error and that the variation in y (effect) is the same for any level of x (cause). If these assumptions do not hold then model II regression is appropriate. Unfortunately the statistical manipulations required are still being developed and there are several techniques that are not wholly satisfactory. Moreover, they are unlikely to be supported by computer packages.

One suggested model II regression is the **Bartlett three group method**. This method simply arranges the data by the magnitude of the x values and divides them into three groups, making sure that there are equal numbers of observations in the largest and smallest thirds. The mean values of x and y are calculated for these two groups only and the slope of the line between the two mean points is calculated.

Another model II regression is the **Kendall robust line-fit method** described above.

Polynomial, cubic and quadratic regression

One of the assumptions of standard regression is that the form of the relationship between x and y is a straight line. If this is not true then the first option is to transform either x or y to make the data better fit a straight line. If this does not help then the assumption can be discarded and polynomial regression (of which quadratic regression is a special case) can be employed. The only difference between polynomial and linear regression is that the best-fit line is not straight. The advantage is that a curved line is almost always a better fit, allowing a better prediction of y for each value of x. The disadvantage is that extra parameters have to be included. There is no longer a single 'slope' value but rather two or more factors that have to be applied to x, x^2, x^3, etc. It is usually either a scatterplot or inspection of the residuals from a linear regression that indicates that the straight line is not the best fit and that some sort of curve should be tried.

It should be noted that each time an extra parameter is added to a regression line the best fit should be better than with fewer parameters. It is possible to get an almost perfect fit to a scatter of points if enough parameters are used. However, the question is: will adding an extra parameter be worth it? This is a judgement call, but it can often be helped by considering the biology. It might be possible to consider that a biological process should be best described by a curved line of some kind, but as the line becomes more complicated the biology underlying the relationship also becomes more difficult to justify.

Tests for more than two variables

Tests of association

Most test of association assume that there are only two variables being considered. However it is often the case that for each 'individual' there are three or more observations. If all that is required is to investigate associations in pairs to examine the strength of associations when other variables are accounted for then these tests are appropriate. This section considers briefly **correlation**, **partial correlation** and its non-parametric equivalent the **Kendall partial rank-order correlation**. Once there are several variables for each individual then multivariate analyses (considered in the next chapter) will often be more appropriate.

Correlation

The simplest method used to analyse data sets with more than two observations for each 'individual' or 'site' is to consider all the possible two-way comparisons that could be made. Statistical packages allow this and will happily produce a large matrix of correlation coefficients that can be trawled to find the largest positive and negative numbers for further investigation.

Note: if the package reports a significance value for each of the correlation coefficients then they should be treated with some caution. Remember that the P-value is just the probability that the null hypothesis is true and we usually reject the null hypothesis when $P < 0.05$. This implies that 20 correlations will produce, on average, one P-value that is less than 0.05 even when there is no association at all. Therefore if there are 10 variables for each 'individual' giving 45 possible pairwise correlation coefficients the chances are very high that one or more will report P-values less than 0.05.

There are two commonly used methods for reducing the critical value for P. The **Bonferroni method** simply divides the critical P-value by the number of tests carried out. So three tests with an overall critical P of 0.05 would give a critical P of 0.05/3 for each test. The **Dunn–Sidák** reduces the critical P using the formula $1 - (0.95^{1/k})$, where k is the number of tests that have been carried out (assuming a critical P of 0.05).

See the section on correlation for the mechanics of using the packages (pp. 167–71).

Partial correlation

A partial correlation coefficient gives a measure of the relationship between two variables when one or more other variables have been held constant. A common use of this technique is in morphological analysis when several variables all relate in some way to size and therefore the bivariate correlation matrix technique employed in the previous section merely confirms that all measures are strongly correlated with each other. The partial correlation of two variables when 'size' is held constant will reveal whether they are related in any other way.

Kendall partial rank-order correlation

The Kendall rank correlation coefficient is the non-parametric equivalent of the Pearson product–moment correlation coefficient that can be used in partial correlation analysis. If variables are known to be distributed in a non-normal fashion then this technique should be employed. Unfortunately, this technique is rarely supported in statistical packages.

Cause(s) and effect(s)

If there are more than two variables which can be labelled 'cause' and 'effect' there are a variety of techniques that can be applied to determine more about the relationship between them. Some of these techniques are simple extensions of ANOVA, regression or correlation analysis while others point towards techniques for data exploration that are covered in the next chapter. **Regression** is used when there are two or more similar 'effect' variables matched to the same 'cause'. **Analysis of covariance (ANCOVA)** can be used when two variables are known to be associated and one is used as the dependent variable in an ANOVA analysis. **Multiple regression** is used when there are several 'cause' variables and a single 'effect'. **Stepwise regression** also has several 'cause' variables and one 'effect' but builds the best-fit model in stages. **Path analysis** is more of a data exploration technique that arranges the interrelationships between several 'causes' and 'effects'.

Regression

If regression has been carried out on several different sets of individuals, in different sites or in different years, then it is possible to compare the slopes of these different analyses to see if they differ. Most statistical packages do not support this type of analysis directly although a visual comparison of several analyses can be made by plotting the estimated slopes and their confidence intervals to see if they overlap.

Analysis of covariance (ANCOVA)

This technique is something of a hybrid between analysis of variance and linear regression. Imagine that a field experiment has been set up with plots given one of five different levels of additional CO_2. Data are collected for sap sugar concentration. This appears to be a simple ANOVA-type of design. However, the plots have been surveyed for a range of physical parameters and are known to differ in the amount of organic material in the soil and this affects the sap sugar concentration. The ANCOVA test will effectively use a regression analysis to remove the effect of the organic material level *before* the standard ANOVA is attempted. The factor accounted for by the regression is called the *covariate*, hence the name analysis of covariance. ANCOVA is supported by both MINITAB and SPSS.

Analysis of covariance can be used with any ANOVA design and is a very useful statistic whenever something known to affect the data can be quantified accurately. ANCOVA makes all the assumptions of ANOVA as well as those of regression. It makes the further assumption that all groups have the same linear relationship with the covariate.

Example. A laboratory investigation into the physiological differences between two species of amphipod (shrimp) has shown that although females of both species push water across their developing eggs they do so with different beat frequencies. A further investigation has been made using video cameras in the field to see if this relationship still holds. The investigators also know that beat frequency for both species increases with water temperature and this has been recorded.

Species	Water temperature (°C)	Beats/min
1	10.1	89.0
1	12.2	94.8
1	13.5	99.6
1	11.2	93.8
1	10.2	91.0
1	9.8	89.2
2	14.1	104.5
2	12.3	103.6
2	9.5	91.1
2	11.6	99.6
2	10.1	99.1
2	9.4	88.7

An analysis of variance or *t*-test show that the measure of beats/min is not significantly different between the two species (ANOVA: $F_{1,10} = 2.419$, $P > 0.1$).

However this does not account for the effect of temperature on beat frequency. An analysis of covariance using species as the grouping variable and temperature as a covariate shows that there is a highly significant difference between the two species ($F_{1,9} = 12.436$, $P = 0.006$). Note that analysis of covariance has one fewer degree of freedom for each covariate than is used for an analysis of variance on the same data.

SPSS Input the data in columns as in the example. Label the columns appropriately. From the 'Analyze' menu select 'General Linear Model', then 'Univariate ...'. Move the observed data ('Beats/m' in the example) to the 'Dependent Variable:' box. Move the grouping variable ('Species' in the example) to the 'Fixed Factor(s):' box. Move the covariate ('Temp' in the example) to the 'Covariate(s):' box. Click 'OK' to run the test.

Univariate Analysis of Variance

Between-Subjects Factors

		N
Species	1	6
	2	6

Tests of Between-Subjects Effects

Dependent Variable: Beats / m

Source	Type III Sum of Squares	df	Mean Square	F	Sig.
Corrected Model	312.262a	2	156.131	27.145	.000
Intercept	842.580	1	842.580	146.494	.000
TEMP	241.209	1	241.209	41.937	.000
SPECIES	71.053	1	71.053	12.354	.007
Error	51.765	9	5.752		
Total	109425.360	12			
Corrected Total	364.027	11			

a. R Squared = .858 (Adjusted R Squared = .826)

This confirms the number of observations for each level of the fixed factor ('Species' in the example). The rest of the output is an ANOVA table designed to cope with many factors and therefore has extra lines that appear totally superfluous for this simplest possible scenario. The first column, 'Source', gives the source of the variation. It is the line labelled with the name of the main effect that is most important ('Species' in the example). The other columns give: 'Type III Sum of Squares', then 'd.f.' or degrees of freedom, and then 'Mean Square' (the mean square value is the sum of square value divided by the degrees of freedom). As there are two species there is 1 degree of freedom for 'Species';

the covariate also has 1 degree of freedom. There were six observations within each species giving 11 degrees of freedom for 'Corrected Total'.

Finally comes the important bit: the *F*-ratio, labelled 'F' here. This is the mean square for 'Species' or 'Temp' divided by that for 'Error' (or residual). SPSS gives the *P*-value associated with this value of 'F' and these degrees of freedom and labels it 'Sig.'. In biology we usually look for a value less than 0.05. Here the probability is 0.007 for 'Species' and indicates that the 'Beats/min' for the two species are significantly different from each other once the effect of temperature is accounted for.

MINITAB Analysis of covariance is very simple to carry out in MINITAB. Input the data in columns, exactly as in the example and label appropriately. From the 'Stat' menu select 'ANOVA' then 'General Linear Model...'. Move the observed data ('BPM' in the example) into the 'Responses:' box, the grouping variable ('species' in the example) into the 'Model:' box. Click on the 'Covariates...' button and add the covariate ('temp' in the example) to the 'Covariates:' box. Click 'OK' twice.

(*Or, if the command interface is enabled, type 'GLM c1 = c2;' at the MTB > prompt and 'covariates c3.' at the SUBC > prompt, assuming that the data is in c1, the main effect is in c2 and the covariate is in c3.*)

General Linear Model: BPM versus Species

```
Factor      Type Levels Values
Species     fixed    2 1 2

Analysis of Variance for BPM, using Adjusted SS for Tests

Source    DF    Seq SS    Adj SS    Adj MS      F      P
Temp       1    241.21    241.21    241.21  41.94  0.000
Species    1     71.05     71.05     71.05  12.35  0.007
Error      9     51.76     51.76      5.75
Total     11    364.03

Term          Coef   SE Coef       T      P
Constant    62.302     5.147   12.10  0.000
Temp        2.9580    0.4568    6.48  0.000

Unusual Observations for BPM

Obs       BPM       Fit   SE Fit  Residual   St Resid
 11    99.100    94.611    1.094     4.489      2.10R

R denotes an observation with a large standardized residual.
```

The output first confirms that the test is a general linear model, it then lists the factor(s) (grouping variables) giving the number of groups (labelled 'Levels') and the codes assigned to the groups (labelled 'Values').

Then comes another confirmation of the test followed by a simple ANOVA table with an extra row for the covariate. The table has the degrees of freedom ('D.F.'), two versions of the sum of squares ('Seq SS' and 'Adj SS'), the mean square ('Adj MS' which is SS/d.f.), the *F*-ratio ('F' which is factor MS/error MS) and finally the *P*-value ('P'). If the *P*-value is less than 0.05 then the null hypothesis is rejected. In the example the effect of temperature as a covariate is confirmed to be highly significant: 'P' is given as '0.000' which should be reported as $P < 0.001$. The factor 'Species' has a *P*-value of 0.007 so the null hypothesis that species have the same 'BPM' is rejected and the alternative hypothesis that the two species differ is accepted.

Finally, there is a regression style analysis of the covariate ('Temp' in the example). It shows that the relationship has a positive slope ('Coef' can be thought of as the slope of the relationship and is 2.958 in the example). The probability that the slope is 0 is very small, given as '0.000' in the example or $P < 0.001$.

Excel There is no way of carrying out ANCOVA.

Multiple regression

If there are several 'cause' variables set, or chosen, by the experimenter and a single 'effect' variable then multiple regression may be appropriate. The same assumptions as for linear regression apply so each of the 'cause' variables must be measured without error. There is an additional assumption that each of the 'cause' variables must be independent of each other. Multiple regression works in exactly the same way as linear regression only the best-fit line is made up of a separate 'slope' for each of the 'cause' variables. There is still a single 'intercept', which is the value of the 'effect' variable when all the 'cause' variables are zero.

A multiple regression using just two 'predictor' or 'cause' variables can be visualized using a 3D scatterplot, but if there are any more 'predictors' there is no way to satisfactorily display the relationship.

The technique of multiple regression is rather overused as it is supported, and fairly easily accessible, in most statistical packages. There is an implicit assumption that all the 'cause' variables impinge directly on the 'effect' variable, which may appear uncomfortable or unreasonable in certain circumstances. Multiple regression can be found in both SPSS and MINITAB.

Stepwise regression

The assumptions and conditions of stepwise regression are identical to multiple regression. The difference is in the way the best-fit model is generated. In stepwise regression the 'causes' are added and subtracted in steps only using those combinations and slopes that generate a better fit (i.e. smaller residuals). This technique is useful as it will help to identify those 'cause' variables that are most important, which can lead to better experimental design in the future.

There are three strategies for building a regression model from a set of possible predictor variables. The simplest is the *forwards* strategy. This picks the best of the predictors and then adds the next best and so on, until adding a further variable does not increase the prediction power of the analysis further. The opposite strategy is *backwards* where the analysis starts with all the variables used and then removes the one that has the smallest impact on the predictive power and so on.

The problem with these strategies is that once a variable is added (or subtracted) it remains in (or out) for the rest of the analysis. *Stepwise* regression builds a model in the same way as the *forward* strategy, but each time a variable has been added or subtracted all variables are then considered for addition or removal. In this way, the best fit from any combination of the variables will be made.

The problem for the user of stepwise regression is that decisions have to be made before the test is run. The user must decide how much of an improvement to the prediction is required for a variable to be worth adding to the model and how little loss in the predictive power is allowed for a variable to be dropped.

Stepwise regression is not guaranteed to find the best possible prediction of the 'effect' from the 'cause' variables available. The only way to do this is to try every possible combination of 'causes' and choose the best one. This strategy was once far too time-consuming due to the number of calculations required, but, with ever faster computing, it is sometimes offered as an option in statistical packages.

Path analysis

This is a technique related to regression and correlation that removes much of the 'cause' and 'effect' baggage. The idea is to generate a map of the interrelationships between variables to visualize the effects and associations within groups of variables. It can be considered as an extension to stepwise regression with multiple possible 'effect' variables. The main difference between path analysis and other techniques with multiple 'causes' is that it allows variables to be correlated, i.e. they do not have to be independent. Unfortunately this analysis is not supported in most statistical packages.

9

The tests 3: tests for data exploration

Data exploration can be attempted with almost any type of data, although in practice it is most useful where there are a large number of variables and observations. The main aim of data exploration techniques is to synthesize and process the interrelationships between observations in such a way to make the patterns obvious to the experimenter. Of all the areas of statistics covered in this book this is the one where there is the most scope for personal choice. These techniques are less concerned with *P*-values and should be treated as ways to generate hypotheses rather than to test them.

Several of the more commonly employed techniques are considered here. This is certainly not an exhaustive list but it does provide a flavour of the sort of techniques that are available.

Observation, inspection and plotting

The simplest and perhaps most obvious way to generate new hypotheses or to explore relationships between variables is to plot them. Many statistical packages, including SPSS and MINITAB will produce a matrix of scatterplots with each cell of the matrix having a different plot of two variables. This sort of visual aid will give a general feel for which variables are related to which as well as for the 'shape' of the data. Do not be afraid to experiment with different types of plot before moving on to more standard methods.

Principal component analysis (PCA) and factor analysis

These are two very similar techniques that weight all the available variables to provide the maximum discrimination between individuals. The idea of principal component analysis (a.k.a. factor analysis, principal axes) is very similar, in many ways, to correlation and regression. The technique can be applied to any data set that has two or more observations for each individual (e.g. several different morphometric measurements from the same specimen). There are assumptions about the data — that it is continuous and normally distributed — but these can be overlooked if the purpose of the test is to generate further hypotheses.

The technique can be visualized well when there are only two observations for each individual. First imagine a scatterplot of two variables that are correlated such that the points fall within an oval cloud. PCA will determine the line through the points that passes through the long axis of the cloud and will use that as the first principal axis or 'principal component'. A line through the cloud of points at right angles to the first axis will generate a second principal component. Of course this process occurs in multidimensional space within the computer with one dimension for each of the variables included in the analysis, and the 'lines' through the clouds of points are formed by weighting each of the variables appropriately.

In this way PCA synthesizes the data from a mass of variables into a set of compound axes. The first axis will explain the most variation, the second will explain the next most variation, and so on. Therefore, inspection of the weightings of the first few axes will show which variables contribute most to the differences between individuals.

In morphometric analysis it is usually the case that individual specimens will vary in size. The first principal axis will nearly always account for size and it is often employed as a method for removing size from the analysis, leaving aspects of 'shape' for the second and subsequent axes.

Example. A typical use of PCA and factor analysis is for exploration of morphometric characters. Here there is a very small data set from only 16 individuals of two species of fruit fly, *Drosophila melanogaster*, and *D. simulans*. Five morphological characters have been measured to the nearest 0.01 mm; the sex and species is recorded too. The species are coded '1' and '2' for convenience and sex is coded '1' for female and '2' for male.

Sex	Species	Thorax length (mm)	Wing length (mm)	Femur length (mm)	Eye width (mm)	Third anterior segment (mm)
1	1	1.01	2.51	0.06	0.52	0.11
1	1	0.98	2.45	0.05	0.53	0.12
1	1	1.02	2.57	0.08	0.55	0.11
1	1	1.05	2.61	0.07	0.52	0.10
2	1	0.98	2.40	0.04	0.54	0.13
2	1	0.89	2.35	0.04	0.50	0.14
2	1	0.89	2.38	0.05	0.50	0.12
2	1	0.95	2.41	0.05	0.49	0.12
1	2	1.20	3.10	0.09	0.48	0.09
1	2	1.15	3.12	0.10	0.52	0.10
1	2	1.18	3.21	0.09	0.52	0.11
1	2	1.21	3.20	0.10	0.55	0.09
2	2	0.95	2.51	0.08	0.56	0.11
2	2	0.94	2.50	0.07	0.49	0.13
2	2	0.96	2.62	0.08	0.51	0.14
2	2	0.91	2.45	0.07	0.52	0.13

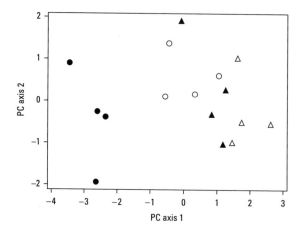

Fig. 9.1 Plotting individuals according to their principal components in MINITAB. PCA maximizes the difference between individuals rather than groups. The individuals are coded by species (shapes) and sex (filled or open shapes). The filled circles are clearly different to the other groups on axis 1 but there is no obvious pattern on axis 2. In many PCA analyses on morphological data axis 1 is closely related to size and other axes to 'shape'.

It is important to realize that there is no requirement for the grouping variables 'sex' and 'species' as the PCA operates on the measured variables to maximize the differences between individuals (i.e. the rows in this data set). I have used the coded variables only to illustrate the results of the analysis.

There are three main components to the output. The first is the weighting applied to each of the variables to generate the principal axes. The second is the set of eigenvalues that show how important the principal axes are (each axis will always explain less of the variation than the last). The third is the position of the individuals on the axes; it is this that is used to generate the graphical display of the output (Fig. 9.1).

Investigation of the factor weightings will show which characters are being used to generate the differences between individuals and which are not (the ones with weightings near zero). In the example the first axis (using either PCA or factor analysis) is generated by contrasting the three size variables 'wing', 'thorax' and 'femur' against 'third anterior segment'. This allows the investigator to focus on these characters as the most important.

The position of individuals on the axes can be used in ANOVA to see if the groups vary. In the example a two-way ANOVA on the first principal component (PC1) using 'sex' and 'species' as grouping variables proved that both factors were highly significant and that they had a significant interaction too. A similar test on PC2 showed no significant differences at all. This shows that there are morphological differences between the sexes and species that are condensed into the first axis.

SPSS Input the data in the same form as the table in the example. Label the columns appropriately. From the 'Analyze' menu, select 'Data reduction', then 'Factor...'. In the dialogue box highlight all the measured variables (but not any grouping variables) and move them into the 'Variables:' box. You can click 'OK' now although there are two detours that may pay dividends. Clicking on 'Extraction' will allow you to determine the number of axes ('Factors:') that are generated (I clicked on 'Number of factors:' and selected two for this example). Clicking on 'Scores' allows you to store the scores as variables. If you do this then you can generate a figure similar to the one that I have generated from the example data by using a scatterplot.

The output generated is as follows:

Factor Analysis

Communalities

	Initial	Extraction
THORAX	1.000	.932
WING	1.000	.939
FEMUR	1.000	.794
EYE	1.000	.994
THIRD_AN	1.000	.749

Extraction Method: Principal Component Analysis.

Total Variance Explained

Component	Initial Eigenvalues			Extraction Sums of Squared Loadings		
	Total	% of Variance	Cumulative %	Total	% of Variance	Cumulative %
1	3.419	68.386	68.386	3.419	68.386	68.386
2	.988	19.765	88.152	.988	19.765	88.152
3	.383	7.658	95.810			
4	.193	3.857	99.666			
5	1.668E-02	.334	100.000			

Extraction Method: Principal Component Analysis.

Component Matrix^a

	Component	
	1	2
THORAX	.959	-.112
WING	.952	-.178
FEMUR	.891	1.764E-02
EYE	.229	.970
THIRD_AN	-.864	-4.63E-02

Extraction Method: Principal Component Analysis.
a. 2 components extracted.

Of the three output tables, the second and last are useful. In the 'Total Variance Explained' table the summaries for the extracted principal component axes are given. Five 'Components' appear on the left of the table and only two are duplicated on the right, because I chose to have only two factors generated. The maximum number of components, factors or axes will be the same as the number of variables in the data set (five in the example). The first 'Component' has the largest eigenvalue and explains the most variance. In the example an eigenvalue of 3.419 translates as nearly 69% of the variation between individuals. The first two 'Components' combined explain 88% ('Cumulative %') of the variation between individuals between them.

The last table gives the 'Component Matrix' and is a summary of the weightings assigned to each of the variables for the extracted components or factors. In this case factor 1 ('Component 1') shows a high positive weighting for the first three variables contrasted against a high negative weighting for 'Third_an', whereas 'Component 2' gives a high positive weight to 'Eye'. This suggests that the best contrast between individuals is achieved by comparing the size variables with the size of 'Third_an'. Perhaps individuals with large values for 'Thorax', 'Wing' and 'Femur' tend to have small 'Third_an' and vice versa.

Some of the weightings are given in unhelpful scientific notation. The weighting of '1.764E-02' translates as 1.764×10^{-2} or 0.01764, a very low weighting.

Two new variables will have appeared in the 'Data View', called 'fac1_1' and 'fac2_1'. These hold the weighted scores for the first two axes and a scatter of these two variables will allow the generation of a figure such as Fig. 9.1.

MINITAB Input the data in columns as set out in the example. Label all the columns appropriately. From the 'Stat' menu choose 'Multivariate' and then 'Principal components...'. In the dialogue box highlight all the measured variables, then click on 'Select'. The analysis will calculate as many components as there are variables (i.e. five in the example) unless you choose a lower number in the 'Number of components to compute:' box. If you wish to keep the position of individuals on each of the axes then you must choose columns for values in the 'Scores:' box. In the example I selected two components and then stored them in columns c9 and c10 and then scatterplotted c9 and c10 to produce the figure.

If you click 'OK' this is the output you will get:

Principal Component Analysis: thorax lengt, wing length , femur length, eye widt

Eigenanalysis of the Correlation Matrix

Eigenvalue	3.4193	0.9883	0.3829	0.1928	0.0167
Proportion	0.684	0.198	0.077	0.039	0.003
Cumulative	0.684	0.882	0.958	0.997	1.000

Variable	PC1	PC2
thorax l	-0.519	-0.113
wing len	-0.515	-0.179
femur le	-0.482	0.018
eye widt	-0.124	0.976
third an	0.467	-0.047

The output confirms the test. Then it gives the eigenvalues of the principal components from one to five. The eigenvalue can be treated as a measure of the variation explained by the axis. Each factor will explain less of the variation than the previous one. This is converted to a proportion of the variation and cumulative proportion for convenience. So the first axis explains 68.4% (0.684) of the variation and the second 19.8%.

In the second part of the output the weightings applied to the five measured variables are given. The first principal component can be interpreted as a contrast between 'third an' and 'thorax', 'wing' and 'femur', while the second is heavily influenced by 'eye'. These weightings maximize the differences between individuals in the sample.

Excel It is not possible to carry out this test in Excel.

Canonical variate analysis (CVA)

This technique works in very much the same way as PCA but with one crucial difference — the individuals must be assigned to groups before the analysis is run. The test then calculates the variable weightings that will maximize the differences between *groups* rather than individuals as is the case with PCA.

Canonical variate analysis produces weightings that will allow you to identify those variables that are the most different between groups and discard those that are the same. It is important to use real classifications rather than arbitrary ones to get the maximum benefit from the technique.

Discriminant function analysis

As with CVA, this technique also requires that the individuals be divided into groups. The idea of discriminant function analysis is to provide a set of weightings that allow the groups to be distinguished. The weightings can then be used on individuals that are not assigned to a group to provide a probability of them belonging to each of the possible groups. If the probability is high then the 'unknown' can be confidently assigned to a group. The power of the weightings is often tested by removing a portion (one or more observation) from the data set, using the remainder to create the weightings and then to use the weightings to assign the removed individuals to groups. The hit rate is a measure of the power of the test to discriminate real unknowns.

Example. Using the same example as for the PCA but using one of the grouping variables as the target discriminator. The analysis will determine how often an individual of unknown species or sex would be attributed to the correct group.

SPSS Input the data in columns and label appropriately. From the 'Analyze' menu select 'Classify', then 'Discriminant...'. Move the variable with the group codes into the 'Grouping variable:' box. Click on 'Define range...' to input the lowest and highest group code numbers. (In this case I used the variable 'sex' and defined groups 1 and 2.) Click 'Continue' to return to 'Discriminant Analysis'. Next move all the measured variables into the 'Independents:' box.

It is important to visit the 'Classify' dialogue before the test is run. There the 'Prior probabilities' should be changed to 'Compute from group sizes' unless you are sure that unknown individuals should have an equal chance of being in either group (in the example I left 'All groups equal' as I was classifying to sex). Also on the 'Classification' dialogue is the very useful option of 'Leave-one-out classification'. Output from this option is shown below.

There are now a range of possible options that can be selected from the buttons at the bottom. The most useful is 'Save...' where the 'Predicted group membership', 'Probabilities of group membership' and 'Discriminant scores' can be stored as separate variables. If you do not do this then there is no way of checking either which individuals have been assigned to each group or the accuracy rate of the discriminant analysis. Another important option to consider is under the 'Classify...' button where you can either allow the probability of group membership to be equal or to be determined by the relative size of the groups. In the example the groups are of equal size anyway so there is no difference but this will not always be the case. Click 'OK'.

Masses of output appears in the 'Output' window of which this is an edited section:

Standardized Canonical Discriminant Function Coefficients

	Function
	1
THORAX	2.076
WING	-1.659
FEMUR	.170
EYE	.090
THIRD_AN	-.261

Structure Matrix

	Function
	1
THORAX	.770
THIRD_AN	-.744
WING	.523
FEMUR	.370
EYE	.139

Pooled within-groups correlations between discriminating variables and standardized canonical discriminant functions
Variables ordered by absolute size of correlation within function.

Functions at Group Centroids

	Function
SEX	1
1	1.524
2	-1.524

Unstandardized canonical discriminant
functions evaluated at group means

The tables shown are first 'Standardized Coefficients' where the weightings given to each of the measured variables to maximize the differences between groups are given. Clearly 'Thorax' and 'Wing' are the highest weighted in the example. The 'Structure Matrix' gives a slightly different view of the data, as it appears that although 'Thorax' has the highest correlation with the discriminating function the negative correlation for 'Third_an' is almost as important once the relatively small size of the measurements has been accounted for.

Finally comes the 'Group Centroids' for the two groups on the axis of the discriminant function. As there are only two groups and they are of equal size in the example it is not surprising that the group centroids (mean position) are the same distance either side of zero. In practice this means that any individual that has the weightings applied to the measurement variables and scores more than zero will be assigned to group 1, and scores less than zero to group 2.

With the 'Leave-one-out classification' selected the following output also appears:

Classification Statistics

Prior Probabilities for Groups

		Cases Used in Analysis	
SEX	Prior	Unweighted	Weighted
1	.500	8	8.000
2	.500	8	8.000
Total	1.000	16	16.000

Classification Results[b,c]

		SEX	Predicted Group Membership		Total
			1	2	
Original	Count	1	8	0	8
		2	1	7	8
	%	1	100.0	.0	100.0
		2	12.5	87.5	100.0
Cross-validated[a]	Count	1	7	1	8
		2	2	6	8
	%	1	87.5	12.5	100.0
		2	25.0	75.0	100.0

a. Cross validation is done only for those cases in the analysis. In cross validation, each case is classified by the functions derived from all cases other than that case.

b. 93.8% of original grouped cases correctly classified.

c. 81.3% of cross-validated grouped cases correctly classified.

Here the 'Prior Probabilities' confirms that there was a 0.5 probability of being assigned to either group (i.e. 50% female, 50% male). The analysis gives the results of running the analysis once for each individual with the target individual excluded and then reassigned to a group based on the discriminant function score. The 'Classification Results' table shows that all of group 1 (females) were assigned to group 1, but only 7 of 8 in group 2 were assigned to the correct group. This gives an overall efficiency of 93.75% for the discrimination of groups.

The group assignments for each individual are made in a new variable ('dis_1') on the spreadsheet. In the example the one misclassified individual, row 5, should have been in group 2 but was assigned to group 1. It has a 'dis1_1' (discriminant function score) of 0.032 and had a 52% chance of being in group 1 and a 48% chance of being in group 2. All other individuals were assigned to the correct group.

MINITAB Input the data into columns. Make sure that the grouping variable is coded as an integer (e.g. female as '1'; male as '2'). Label the columns appropriately. From the 'Stat' menu select 'Multivariate', then 'Discriminant analysis . . .'. Move the grouping variable ('sex' or 'species' in the example) into the 'Groups:' box. Highlight all the measured variables and click 'Select' to move them into the 'Predictors:' box. Click 'OK'. (In the example I chose 'sex' as the grouping variable as the analysis was 100% accurate when 'species' was used.)

You will get the following output:

```
Discriminant Analysis: sex versus thorax lengt, wing length , ...

Linear Method for Response:    sex
Predictors:  thorax l  wing len  femur le  eye widt  third an

Group            1         2
Count            8         8

Summary of Classification

Put into      ....True Group....
Group            1         2
1                8         1
2                0         7
Total N          8         8
N Correct        8         7
Proportion    1.000     0.875

N =   16      N Correct =    15      Proportion Correct = 0.938

Squared Distance Between Groups
                 1         2
1          0.00000   9.28891
2          9.28891   0.00000

Linear Discriminant Function for Group
                 1         2
Constant      -795.5    -759.6
thorax l       927.6     838.2
wing len      -117.7     -97.2
femur le      -264.2    -293.4
eye widt      1155.6    1143.9
third an      3098.6    3174.4

Summary of Misclassified Observations

Observation      True      Pred    Group    Squared    Probability
                Group     Group             Distance
    5 **           2         1        1       9.183        0.524
                                      2       9.378        0.476
```

The output confirms the test and then the variables used for grouping and prediction. It then gives the number of observations in each of the groups (in the
example, there were 8 females and 8 males). Next comes a 'Summary of
Classification' where the real group ('True group') and the group that each individual is put into are compared. In the example all of group 1 were correctly
assigned to group 1 but one individual from group 2 was incorrectly assigned to
group 1. A summary of the accuracy comes next, giving it as a proportion correct:
0.938 in the example (i.e. 93.8% or 15/16 of the individuals were assigned to the
correct group). The squared difference between the groups is not very useful for
just two groups but will show which groups are most similar to which if there are
more than two. Then comes a section labelled 'Linear Discriminant Function'
which gives weightings to each of the measured variables that are akin to the
slopes in a multiple regression. Finally is a summary of the misclassifications. In
the example only one of the 16 was misclassified. The individual was in row 5.
The analysis gave it a 52.4% probability of being in group 1 (i.e. wrong) and a
47.6% probability of being in group 2 (i.e. correct), so it was a borderline case.

The analysis suggests that an unknown individual from this area could be assigned to 'sex' from an analysis of the measured variables alone with an accuracy of over 93%.

Excel You cannot carry out this test in Excel.

Multivariate analysis of variance (MANOVA)

All the analysis of variance techniques that are introduced in Chapter 7 have only one observed variable although they had one or more classification variables. Therefore they can all be called *univariate* analyses. The statistical analysis MANOVA allows more than one observed variable to be analysed at once, hence is a *multivariate* test. This test often has the effect of combining two or more rather borderline significant results into a highly significant single result.

Example. Using the data for fly morphology again a MANOVA is used on two response variables, 'thorax length' and 'wing length', simultaneously with 'species' being used as the factor. In two univariate ANOVAs on these data 'thorax' is not significant, while 'wing' is significant at $P = 0.01$. The MANOVA uses information from both variables to give a highly significant result of $P = 0.001$.

SPSS From the 'Analyze' menu, select 'General Linear Model' and then 'Multivariate'. Move the observation data to the 'Dependent Variables:' box and the main effect ('Species' in the example) to the 'Fixed Factor(s):' box. Click 'OK'.

General Linear Model

Between-Subjects Factors

		N
SPECIES	1.00	8
	2.00	8

Multivariate Tests[b]

Effect		Value	F	Hypothesis df	Error df	Sig.
Intercept	Pillai's Trace	.992	817.233[a]	2.000	13.000	.000
	Wilks' Lambda	.008	817.233[a]	2.000	13.000	.000
	Hotelling's Trace	125.728	817.233[a]	2.000	13.000	.000
	Roy's Largest Root	125.728	817.233[a]	2.000	13.000	.000
SPECIES	Pillai's Trace	.674	13.447[a]	2.000	13.000	.001
	Wilks' Lambda	.326	13.447[a]	2.000	13.000	.001
	Hotelling's Trace	2.069	13.447[a]	2.000	13.000	.001
	Roy's Largest Root	2.069	13.447[a]	2.000	13.000	.001

a. Exact statistic

b. Design: Intercept+SPECIES

Tests of Between-Subjects Effects

Source	Dependent Variable	Type III Sum of Squares	df	Mean Square	F	Sig.
Corrected Model	THORAX	3.331E-02[a]	1	3.331E-02	3.163	.097
	WING	.574[b]	1	.574	8.933	.010
Intercept	THORAX	16.545	1	16.545	1570.996	.000
	WING	112.307	1	112.307	1748.382	.000
SPECIES	THORAX	3.331E-02	1	3.331E-02	3.163	.097
	WING	.574	1	.574	8.933	.010
Error	THORAX	.147	14	1.053E-02		
	WING	.899	14	6.423E-02		
Total	THORAX	16.725	16			
	WING	113.780	16			
Corrected Total	THORAX	.181	15			
	WING	1.473	15			

a. R Squared = .184 (Adjusted R Squared = .126)

b. R Squared = .390 (Adjusted R Squared = .346)

There are three tables of output. In the first the number of observations in each group is confirmed. In the second output comes the results of the MANOVA. The section on the 'Intercept' is not very informative; it is the bottom half with the effect labelled as 'Species' where the results come. There are four different MANOVA statistics and, in this case, they all give the same results. For all four the *P*-value (labelled 'Sig.') is 0.001 and the *F*-ratios and degrees of freedom are identical too.

In the final table the univariate ANOVAs on the same data are shown. In this case the important line is the one with the source labelled 'Species'. One of the ANOVAs has a *P*-value of 0.097 (i.e. not significant) while the other has $P = 0.010$, which is significant, but not as conclusive as the result from the MANOVA.

MINITAB From the 'Stat' menu, select 'General MANOVA…'. Move two or more variables to the 'Responses:' box. Move the main effect ('species' in the example) to the 'Model:' box. Click 'OK'.

The following output appears:

```
General Linear Model: thorax length (m, wing length (mm) versus species)

MANOVA for species          s = 1    m = 0.0    n =    5.5

Criterion        Test Statistic         F           DF       P
Wilk's              0.32586        13.447  ( 2,    13)  0.001
Lawley-Hotelling    2.06883        13.447  ( 2,    13)  0.001
Pillai's            0.67414        13.447  ( 2,    13)  0.001
Roy's               2.06883
```

This, rather meagre, output can be augmented by requesting various options. A useful option is to run the univariate ANOVAs on all the variables used in the MANOVA to see how much better the groups are separated. In the output there are three different methods of calculating the significance of a MANOVA. As is clear from the output, they all provide the same conclusion with identical 'P', 'F' and 'DF' values. In each case the P-value is 0.001, indicating that the two species are highly significantly different.

Excel There is no method for carrying out this test.

Multivariate analysis of covariance (MANCOVA)

This technique is related to MANOVA in the same way that ANCOVA is related to ANOVA. If there are more than one observed variables, one or more ways of classifying the data and, furthermore, there is a measured observation that is known to have an effect on the observations then MANCOVA can be used to remove the effect of this confounding variable from the analysis.

To carry out this test, follow the MANOVA example and add a covariate to the box indicated.

Cluster analysis

This is a general term for a huge range of techniques for the classification of individuals. These techniques are becoming increasingly important as more detailed statistical analyses of DNA sequences are being attempted. Cluster analysis can be used to generate dendrograms that show putative phylogenetic relationships, or at least divide individuals into groups that might have taxonomic meaning. However, cluster analysis is certainly not restricted to molecular sequence analysis — it has a long history in taxonomy (both cladistics and phenetics) and in community ecology (particularly in vegetation classification or ordination).

In its simplest form cluster analysis can be imagined as a step-by-step process. First the individuals are depicted as a scatter of points. Then the individuals that are closest together are identified and their similarity recorded as the distance between them (this is sometimes called a Mahalanobis distance). The two points are amalgamated into a single point halfway between the two. The next two closest points are then identified and amalgamated. This process continues until there is only one point.

SPSS There are several methods available. As for discriminant function analysis and PCA, input the data in columns so that each individual is represented by a row of observations. From the 'Analyze' menu select 'Classify' and then choose 'Hierarchical Cluster'. Simple dendrograms can be generated by selecting the submenu 'plots' and checking the 'Dendrogram' option. I suggest that for 'Icicle' you select 'None' as this display is very difficult to interpret.

Using the example data and the 'Method' of 'Within-groups linkage' and 'Measure' as 'Pearson correlation', with a 'Rescale to 0–1 range' option I generated the following agglomeration schedule using the example data given above:

Agglomeration Schedule

Stage	Cluster Combined Cluster 1	Cluster Combined Cluster 2	Coefficients	Stage Cluster First Appears Cluster 1	Stage Cluster First Appears Cluster 2	Next Stage
1	10	11	1.000	0	0	7
2	7	16	.998	0	0	8
3	1	4	.988	0	0	6
4	14	15	.984	0	0	13
5	2	5	.976	0	0	11
6	1	3	.969	3	0	9
7	10	12	.967	1	0	12
8	6	7	.965	0	2	10
9	1	8	.957	6	0	11
10	6	13	.913	8	0	13
11	1	2	.912	9	5	14
12	9	10	.899	0	7	15
13	6	14	.883	10	4	14
14	1	6	.764	11	13	15
15	1	9	.512	14	12	0

This shows that the first pair of individuals to be clustered (i.e. the most similar pair) were individuals 10 and 11. These will then form a new group, given the number of the lowest representative in the group, that is compared to the remaining individuals. The next most similar pair were individuals 7 and 16, then 1 and 4 and so on. The group formed in the first step is linked to individual 12 in step 7. This process continues until in step 15 a group of 12 individuals, including individual 1, is linked to a group comprising the remaining four individuals.

The dendrogram shows this graphically:

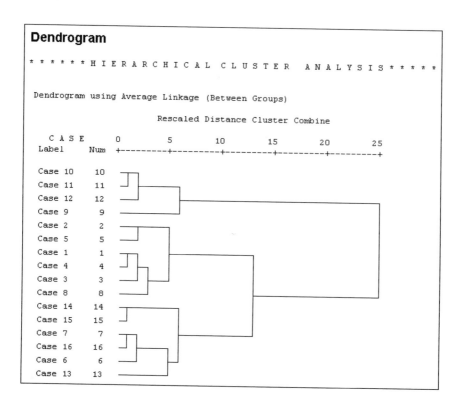

This output is not an editable chart in SPSS and the scale used is rather crude, but it does show graphically the output of the analysis. The most similar individuals are obviously linked (10 and 11, 7 and 16). The individuals 9, 10, 11 and 12 (the females of species 2) clearly form a separate group.

MINITAB There is a variety of clustering methods available in MINITAB. Input the data in columns so that each individual is represented by a single row. From the 'Stat' menu select 'Multivariate' and then one of the three 'Cluster' methods. To generate the dendrogram shown in Fig. 9.2, I used the 'Cluster observations...' method on the measured variables from the example data for PCA.

Excel No clustering techniques are available in Excel.

DECORANA *and* TWINSPAN

DECORANA (detrended correspondence analysis) and TWINSPAN (two-way indicator species analysis) are two analyses developed at the Institute for Terrestrial Ecology (now the part of the Centre for Ecology and Hydrology, UK) that are now widely used in the comparison of communities from species abundance data and in the exploration of ecological data.

Fig. 9.2 A typical dendrogram showing the results of a cluster analysis in MINITAB. This figure was generated using the 16 individuals used as the example data for PCA. There are some clear groupings. For example individuals 9–12 form a separate group on the right of the diagram.

TWINSPAN, as suggested by its name, is very useful for comparing list of species from sets of sites. For example, in a study of woodland carabids the presence and absence data, or measures of abundance, generate a discriminating function that can group similar woodlands and generate dendrograms.

10 Symbols and letters used in statistics

One of the surest ways of making a statistics book difficult to read is the tendency to use Greek letters, single italicized letters or obscure symbols. As far as possible I have tried to avoid this in this book. Here are the ones that you are most likely to encounter.

Greek letters

These are often used to signify the true values of particular statistics (i.e. the value you would get if you were able to measure the entire population rather than a sample). The estimates you get of the true values are often then labelled with the corresponding normal letter.

Π (pi) product of the terms following it (multiply together)
π (pi) a constant (= 3.142) used in geometry
Σ (sigma) sum of the terms following it (add up)
α (alpha) the critical significance level for the rejection of a hypothesis (usually 0.05)
β (beta) true regression coefficient (estimated by the statistic, b)
χ (chi) χ^2 is a commonly encountered statistical distribution
γ (gamma) γ_1 is the true value of skewness; γ_2 is the true value of kurtosis
μ (mu) true mean of a population
ρ (rho) true correlation coefficient (estimated by the statistic, r)
σ (sigma) the true standard deviation of a population
σ^2 (sigma squared) the true variance of a population
T   (tau) the statistic of Kendall's rank order correlation
Δ or δ (delta) increment (tiny difference or change)

Symbols

$-$ (overbar) indicates a mean
$\sqrt{}$ square root
$=$ is equal to
$\equiv$ is identically equal to

$\neq$	is not equal to
$\approx$	is approximately equal to
$\sim$	is distributed as
$\|\ \|$	absolute value of the number between the bars. $\|-6\|=6$
$!$	factorial (e.g. $3! = 1 \times 2 \times 3 = 6$)
$\pm$	plus *or* minus (so 5 ± 2 means a range from 3 to 7)
$\geq$	is greater than or equal to
$<$	is less than (points to smaller value)
$\ll$	is much less than
$>$	is greater than (points to smaller value)
$\gg$	is much greater than
$\wedge$	used in some statistical packages (e.g. Excel) to mean 'raise to the power of'
$\cap; \cup; \subset; \not\subset$	symbols used in set work (intersection; union; is a subset of; is not a subset of)
$*$	indicating a significant result (usually a P-value is flagged at < 0.05)
$*$	used in statistical packages to mean 'multiplied by'
$**$	denotes a highly significant result (usually $P < 0.01$)
$**$	used in some statistical packages (e.g. SPSS) to mean 'raise to the power of'
$***$	denotes a very highly significant result (usually $P < 0.001$)
_____	used to underline groups that are not significantly different (see *post hoc* tests, p. 116)
∞	infinity (an infinite number) often used in statistical tables to indicate the value of an asymptote
$\propto$	varies directly as, is proportional to
$\Rightarrow$	implies
$\times$ or $\cdot$	multiply
$\therefore$	therefore

Upper case letters

CI confidence interval
CL confidence limit
CV coefficient of variation
DF degrees of freedom (also d.f. or df)
F F-value (e.g. the output from ANOVA) ratio of within- and between-group variance
F sometimes used to indicate a function
H_0 null hypothesis (the uninteresting hypothesis — nothing is happening)
H_1 alternative hypothesis (the interesting hypothesis — something is happening)
MS mean square (SS/d.f. in an ANOVA table)

P   probability (more usually p or sometimes P)
PI   prediction interval
SD   standard deviation (also s)
SE   standard error
SS   sum of squares
X^2 estimate of value for χ^2 (chi-square)

Lower case letters

a the intercept of a regression line (where the line crosses the y-axis)
a.k.a.   also known as, not statistics but it is used several times in this book
b slope of a regression line
d.f.   degrees of freedom (sometimes df or DF)
e a constant ($= 2.172$) used as the base for natural or Naperian logarithms (ln)
g estimate of value of γ (gamma): $g_1 =$ skew, $g_2 =$ kurtosis
f used to indicate a function
i often used to indicate a sequence of observations (e.g. x_i)
j often used to indicate a second sequence of observations (e.g. x_{ij})
m often used to indicate the sample mean
p   probability (also P, P or p)
p binomial probability (e.g. 0.5 probability of an individual being female)
r measure of correlation (Pearson product–moment correlation, varies from -1 to 1)
r_s measure of correlation produced by Spearman rank-order correlation
r^2 a measure of the amount of variation accounted for by a regression line or correlation
s standard deviation of a sample (also SD)
s^2 variance of a sample
t value of the statistic resulting from a Student t-test
v occasionally used to indicate variance of a sample
x often used to indicate an observation
y often used to indicate a second observation on the same individual as x
z often used to indicate a third observation on the same individual as x and y

Glossary

This gives an explanation of commonly encountered words, concepts and acronyms including thumbnail description of many statistical tests.

A *posteriori* a phrase applied when a hypothesis is generated after the data have been collected. Sometimes used as a synonym for a *post hoc* test

A *priori* a phrase applied when a hypothesis is generated before the data have been collected

Abscissa the x or horizontal axis of a graph

Accuracy the closeness of a measure to its true value (different to precision)

Alpha the name usually given to the critical value of P required to reject a null hypothesis, usually 0.05

Anderson–Darling test a statistical test used to determine whether a set of data are normally distributed

Angular transformation a synonym of arcsin transform. A transformation that is often used to 'normalize' percentage or proportion data

ANCOVA a short term for analysis of covariance

ANOVA a short term for analysis of variance, the term was originally coined by Tukey

ANOVAR an alternative term for analysis of variance

Arcsin square root transformation *see* arcsin transform

Arcsin transform an operation that is often used to 'normalize' percentage data (synonym for arcsin square-root transformation and angular transformation)

Arithmetic mean a measure of position: the sum of all values divided by the number of observations (synonym of mean)

Association a measure of the strength of the relationship between two variables; often used as a synonym of correlation

Assumptions many statistical tests assume that the underlying distribution of the measurements of a whole population is of a particular type (usually that it follows a normal distribution) or that measurements are made without error, etc.

Asymptote the value of a line that a curve is approaching but never meets

Attribute when used to describe an observation it usually implies a small number of possible categories that have no meaningful sequence (e.g. 'blue', 'pink' and 'white')

Autocorrelation a common problem with sets of observations is that they are not truly random and are more similar to their 'neighbours' in space or time than is expected by chance

Average a synonym for arithmetic mean (used by Excel for arithmetic mean)

Back transformation process by which observations are transformed back to their original units

Balanced design an experiment where the same number of observations are made for each factor level (or factor combination in two-way ANOVA)

Bar chart a graph to display the distribution of a discrete variable where each category on the x-axis represents one possible value

Bartlett test a test for homogeneity of variance

Bartlett three group method a model II regression technique

Bayesian statistics a whole branch of statistics based on a different approach to the concepts of probability and certainty

Bernoulli distribution values can only be one of two possibilities (e.g. 0 or 1)

Best fit the statistical model which explains the most variation. The best fit is often constrained by the type of statistical test. For example, if the model is a linear regression then the 'best fit' will be the straight line which accounts for the most variation (leaves the smallest residual variation)

Between-sample variance a synonym for between-groups sum of squares. In ANOVA and regression it is a measure of the variation between factor levels. Comparison of the between-sample variance and within-sample variance is how ANOVA works

Bias any systematic error in measurement

Bimodal a frequency distribution that has two peaks

Binomial distribution a theoretical probability distribution in events that can occur in two categories. Can be used as a null hypothesis to determine whether given data are random or not

Bivariate tests that are applied to two variables

Bivariate normal an assumption of many bivariate parametric tests is that the two variables are both normally distributed

Bonferroni method a *post hoc* test, used after a one-way ANOVA to determine which groups are different from which. More generally a method for reducing the critical value of alpha (normally 0.05) when many tests are carried out in the same experiment: divide 0.05 by the number of tests to get new value of alpha

Bootstrapping a method of analysis that improves many estimated statistics by using repeated subsamples of a data set

Box and whisker a method of displaying data where a horizontal line represents the median, a box extends to cover the interquartile range and a line may extend away from the box to the extreme values

Box plot synonym of box and whisker

Calibration curve regression can predict y from any value of x. This can be used to generate a calibration curve

Canonical variate analysis a multivariate test where the data are divided into groups and the test weights the variables to maximize differences between the groups

Caption a small piece of text that should accompany *every* figure and table used in a report (synonym for legend)

Categorical a description of data or a variable that is only described by categories. This can be as simple as 'red, blue, green' or an observer-imposed scale, e.g. of aggressiveness where 5 is aggressive and 1 is passive. Categories are often called attributes

Causation the assumption in statistical tests that the value of the independent variable causes the result in the dependent variable

Chi-square distribution a distribution widely used in statistics that is based on the deviation of the sample variance from the true variance of a population. A χ^2 distribution can be generated from a population of standard normal deviates as the probability density of a very large number of samples scaled thus: $((n-1)s^2)/\sigma^2$

Chi-square goodness of fit a chi-square test where the expected values are taken from a particular distribution against which the observed distribution is being compared

Chi-square test a contingency table-based statistical test to explore hypotheses of association between variables; expected values are generated by the table

Cluster analysis a number of multivariate tests that group observations by similarity and may provide insight into the data

Component a synonym for a factor or axis in a principal components analysis

Confidence interval (CI) a measure of spread: it uses the standard error and the t-distribution to give a range of values within which there is a percentage probability of the true mean occurring (usually using set at 95%). In regression the 95% confidence intervals of slope and intercept are used to generate 95% confidence intervals of the line. This zone is not parallel with the best-fit line but is wider at extreme values of x

Confounded design this often occurs in a multifactorial design when not all the combinations have been used or when the effects of one factor cannot be disentangled from the effects of another (i.e. they are not independent). Can be due to lack of space or resources, or a poor experimental design

Conservative test one where the chance of a type I error is reduced and that of a type II error increased (i.e. the null hypothesis is rejected less often than it should be)

Constant any fixed value. In regression analysis the intercept (value of y when x is zero) is sometimes called this

Contingency table a way of displaying data that has been assigned to categories for two variables (e.g. broadleaf/conifer and insect damage > 10%? yes/no)

Continuity correction a method of correcting bias in various contingency tests. Makes test result more conservative

Continuous a description of data or variables indicating every value is possible (in theory); e.g. any linear measurement, > 29 possible values as a rule of thumb

Control a general term given to the section of an experiment that is unmodified by the experimenter

Cook distance statistic a useful measure of the influence of single measures on the outcome of regression analysis (any value > 1 is large)

Correlation a method of measuring the association between two variables (often used as a synonym for the Pearson product–moment correlation). *Warning*: large samples will nearly always produce mathematically 'significant' results even when the biological significance is only miniscule

Covariance technically, this is the sum of the product of the deviations from the mean for a set of paired observations (i.e. the variance that remains once the relationship between two variables is accounted for). It is related to variance and is useful in many statistics

Cox regression a type of regression used for assessing hazards (e.g. mortalities) to make predictions about the chance of events occurring

Cramér coefficient the result of a test on contingency tables other than 2×2

Cramér's V a statistic correcting chi-square values for sample size

Cumulative frequency the number of times an observation takes a particular value *or less* in a data set (e.g. if there were 26 zeroes, 22 ones and 15 twos then the cumulative frequency for two would be $26 + 22 + 15 = 63$)

Cumulative probability the probability of achieving a particular value and all values greater (or smaller). A type of data display where probabilities are accumulated from left to right until all observed values have been included

Data observations recorded during an experiment or survey (plural)

Datum a single observation

Degrees of freedom a number related to the sample size that accounts for the number of observations made, number of factor levels and any manipulations carried out. Degrees of freedom are lost, for example, when a set of values is constrained to total 1 or when a set of data is being compared to a standard distribution with a mean or other parameters the same as the data

Dependent variable the observed or 'effect' variable not set by the experimenter

Derived a value that results from a combination of two or more values (e.g. ratio, proportion or percentage)

Descriptive statistic anything that summarizes the data, e.g. mean or standard deviation

Dichotomous variable a variable that can only take two possible values (e.g. yes/no)

Discontinuous data only a limited number of values are possible (< 30 as a rule of thumb)

Discontinuous variable a variable that only has a limited number of values (usually but not always integers)

Discrete a description of data or variables indicating that not every value is possible, e.g. number of children can never be 2.5

Discriminant function analysis a multivariate test that assigns individuals to groups

Dispersion the way in which the data are distributed, often measured by standard deviation (synonym of spread)

Distribution the spread of a set of observations

Double blind an experiment when the experimenter does not know which treatment is being applied until labels are decoded. A very good way to avoid inadvertent bias by recorders and a requirement of all experiments on humans

Dummy data made up before the experiment is carried out and used for a dry run of the statistical tests planned

Duncan method a *post hoc* test, used after a one-way ANOVA to determine which groups are different from which

Dunn–Sidák method a *post hoc* test, used after a one-way ANOVA to determine which groups are different from which. More generally it is a method of setting the critical value of alpha (usually set to 0.05) when many experiments or tests are carried out at the same time. This method is often considered superior to the Bonferroni method (q.v.); the new alpha equals $1 - (0.95^{1/k})$, where k is the number of tests that have been carried out

Dunn test one of many *post hoc* tests used in ANOVA to determine which groups are different from which

Error a word quite widely used in statistics. Can be applied to the degree of accuracy with which measurements are made. Can also be used in statistical tests such as ANOVA to refer to variation not accounted for by the model being tested

Estimate any statistics (e.g. mean) calculated from a sample is an estimate of that value for the whole population

Expected frequency the occurrence of an event determined from a theoretical distribution or hypothesis based on previous observations

Expected value values needed in contingency tables when they are calculated using the row and column totals or values derived from a theoretical distribution or hypothesis

Experiment a random assignment of subjects to controlled 'experimental' conditions

Experimental design a process of planning which should always occur before an experiment begins to maximize the usefulness of results obtained while minimizing the effort

Exponential distribution a distribution that arises when there is a constant probability of an event occurring (e.g. birth rate or radioactive decay rate). Logarithmic transformations will turn exponential curves into straight lines

Extrapolation whenever predictions are made beyond the range of the data available

F-distribution an asymmetric, continuous distribution, used in ANOVA tests with a modal value of 1, representing the frequency of occurrence of the ratio of between-group variance to within-group variance for groups with the same distribution. Called F in honour of R.A. Fisher

F-ratio a synonym for F-value

F-test a statistical test with the null hypothesis that groups have the same variance. Often offered as a test for homogeneity of variance in a t-test

F-value the output from an ANOVA test. The ratio of between-group variance to within-group variance

Factor a grouping variable or independent variable. For ANOVA, factors must be discontinuous/categorical or made to appear so

Factor level a number representing different groups, e.g. sap flows are examined from three species of tree. The species are groups that are arbitrarily assigned numbers as factor levels

Factorial (experiment) an experiment or survey where there are two or more factors and each possible combination of factor levels is represented in the data

Field experiment a general term given to any experiment where conditions are manipulated by the experimenter and which occurs outside a controlled laboratory

Figure any graphical item in a report

First-order interaction an interaction between two factors in a factorial ANOVA

Fisher's exact test a statistic for 2×2 contingency tables used when the total number of observations is small

Fisher's z-transformation a method for transforming the Pearson correlation coefficient to make it more amenable to analysis

Fixed effect this is often applied to describe a factor in ANOVA that is set by the experimenter (in contrast to a random effect)

Four-way . . . *see* entry for two-way . . . and extrapolate

Frequency the number of times a value (or range of values) occurs in a sample (e.g. there were 56 white-eyed flies in the vial)

Frequency distribution the number of times each possible value occurs in a sample

Friedman test a non-parametric test that is a restrictive version of a two-way ANOVA. Only one observation is allowed for each factor combination

***G*-test** a form of contingency table test (a.k.a. log-likelihood ratio test)

Gamma distribution a family of continuous distributions

General linear model (GLM) a term used to describe a family of analyses that include ANOVA and linear regression

Generalized linear model a synonym for general linear model

Genstat a statistical package (capable of carrying out complicated ANOVA designs)

Geometric mean a measure of position: the antilog of the mean of the logs of each value

Geometric mean regression a model II regression method

GLIM a statistical package very capable in general linear models

Goodness of fit any statistic that compares the actual distribution of a variable with a theoretical one

Graphical representation any method of displaying data visually

Group a set of observations with something in common, e.g. all from the same tree

Grouping variable a variable by which observations can be divided into groups, e.g. tree number

Harmonic mean a measure of position: the reciprocal of the mean of the reciprocals of each value

Heteroscedasticity when different groups have unequal variance and therefore the data violates the assumptions of ANOVA and regression

Hierarchical ANOVA a synonym of nested ANOVA

Highly significant where the null hypothesis is rejected because the P-value is much less than the 0.05 level. The actual level varies with authors but P must be less than 0.01

Histogram a graph to display the distribution of a continuous variable where each category on the x-axis represents a range of values

Homogeneity of variances an assumption of many parametric tests such as the t-test or ANOVA is that the variances in the groups are equal. This can be tested using the Levene test

Homoscedasticity when different groups have equal variance and therefore conform to the one of the assumptions of ANOVA and regression

Hotelling trace a method for calculating P-values in MANOVA

Hypothesis this is what is being tested when a statistical test is carried out, e.g. 'male and female tree frogs have a different mean weight'

Independence an assumption of many statistical tests. The data collected in a sample is not affected by whether another event has occurred. Many statistical tests are rendered invalid because of lack of independence

Independent samples no individuals measured in sample A are also measured in sample B. If this is violated, the sin is called pseudoreplication

Independent samples *t*-test a *t*-test comparing two sets of data that are not paired

Independent variable the 'cause' variable set or varied by the experimenter

Individual in statistics an 'individual' can be a pair, bone, region, species or any number of different things. The 'individual' will provide one observation for each variable under consideration and when the data are arranged in the package will provide a single row

Integer a whole number

Interaction in ANOVA this is a measure of whether two or more grouping variables have an additive (no interaction) effect or not (interaction)

Intercept used in regression analysis: the point on the *y*-axis when the value of *x* is zero

Interquartile range a measure of spread: if the data are in rank order it uses the range from the value 25% down the list to that 75% down

Interval data when observations are made on a meaningful measurement scale

Jack-knifing a technique to determine bias in statistics by recalculating using a subset of the data

Kendall rank correlation a non-parametric test to measure correlation (association)

Kendall rank-order correlation a non-parametric test to measure correlation (association)

Kendall robust line fit a non-parametric model II version of regression (rarely used)

Kolmorogov–Smirnoff (K–S) test a non-parametric test that compares two distributions. Very useful for goodness of fit tests and more powerful and convenient than a chi-square goodness of fit when samples are large

Kruskal–Wallis (KW) test a non-parametric version of a one-way ANOVA that tests the null hypothesis that two or more groups come from populations with the same median

Kurtosis a measure of the shape of a distribution (sometimes called g_2)

Latin square an experimental set-up that reduces possible bias caused by unquantified variation in the environment across an experiment

Lawley–Hotelling test a method for calculating *P*-values in MANOVA

Least significant difference (LSD) test the simplest and probably the most widely used *post hoc* test used after a one-way ANOVA to determine which groups differ from which

Legend a small section of text that should accompany every figure or table making it interpretable without recourse to the main text (synonym for caption)

Leptokurtic a distribution that has a lot of observations around the mean and in the tails, but fewer in the 'shoulders'

Level a particular treatment; a defined condition for the independent variable (set by the experimenter)

Levene test a test for homogeneity of variances. Used for checking data to see if a parametric test such as ANOVA is appropriate

Liberal test a statistical test where the null hypothesis is rejected more often than it should be

Line chart (graph) a graphical representation of data where the points are joined. This makes the assumption that values between points are possible and likely to be fairly represented by the position of the line

Linear regression a regression that assumes the 'cause and effect' relationship is best described by a straight line

Log-likelihood test a method of calculating the probability that a particular contingency table will arise. Basis of the G-test

Log transformation an operation where each observation is logged (usually to base 10)

Logistic curve a distribution, usually with a range from 0 to 1, where the values near the extremes are approached with asymptotes

Logistic regression a version of regression where the 'effect' variable is transformed using logits. Useful when using regression to predict values that have a restricted range of possible values (e.g. proportions or percentages)

Logit transformation used to convert values on a scale with limits (e.g. proportions have limits at 0 and 1) to a limitless one. Transformation for a proportion, x: Logit $x = \ln (x/(1 - x))$

LSD test *see* least significant difference test

Main effect in ANOVA a factor (grouping variable) that is at the top of a design (i.e. not nested)

MANCOVA a MANOVA that includes a covariate

Mann–Whitney U test a non-parametric test of a null hypothesis that two groups come from the same distribution. Synonym of the Wilcoxon–Mann–Whitney test and the Wilcoxon signed ranks test

MANOVA an acronym for multivariate analysis of variance. A test where there is more than one dependent variable under investigation

Mantel test a matrix method for determining how two matrices of data are associated

Matched samples or data a synonym for paired samples or repeated measures

Mean a measure of position: sum of all values divided by the number of observations (synonym of arithmetic mean and average)

Mean square (MS) used in ANOVA; the total sum of squares divided by the degrees of freedom (SS/d.f.)

Measured variable a variable where the experimenter has to take a reading, e.g. height, weight, optical transmission

Measurement an observation or single item of data (datum)

Median a measure of position: if all the data are put in rank order it is the value of the datum in the middle

Meta-analysis a method for combining the results of several tests of the same hypothesis. It can be used even when the original data are not available and different tests have been used

MINITAB a statistical package

Mixed model (ANOVA) a test with both fixed and random grouping variables (factors)

Mode a crude measure of position: the most commonly occurring value

Model I ANOVA the basic ANOVA where the grouping variables (factors) are all fixed effects

Model II ANOVA an ANOVA where all the grouping variables (factors) are random effects

Model I regression the usual parametric regressions that assumes the 'cause' variable is measured without error

Model II regression a rarely used version of regression that takes into account the fact that the 'cause' variable may be measured with error

Multifactorial design there are many variables that can be used to group the data

Multiple correlation a difficult to interpret comparison of more than two variables

Multiple regression a test that establishes the best prediction of an 'effect' variable using all 'cause' variables simultaneously

Multivariate statistics tests which use more than one dependent variable

Negative binomial distribution a discrete probability distribution which is frequently invoked to describe contagious (clumped) distributions

Nested ANOVA a synonym of hierarchical ANOVA. A test where at least one of the grouping variables is a subgroup of another (e.g. 'bunch' as a grouping variable within the grouping variable 'vine' in an experiment on grapes)

Nominal where the values in a data set cannot be put into any meaningful sequence, only assigned to categories (e.g. blue and red)

Non-parametric test a test where few or no assumptions about the shape of a distribution are made

Normal distribution a unimodal, continuous probability distribution with a characteristic bell-shaped curve. This distribution is often assumed of the data in parametric statistics

Null hypothesis every hypothesis being tested must have a null hypothesis, e.g. if the hypothesis is that two groups have different mean heights then the null hypothesis must be that the two groups do not have different mean heights

Observation a single item of data (datum), a measurement

Observer bias whenever two, or more, observers have differences in the values recorded from the same observations. Can often be a consistent difference that can be corrected for

One-tailed test a test that assumes rejection of the null hypothesis can only come from a deviation in one direction rather than either. For example, the hypothesis that two groups are different is 'two-tailed' while the hypothesis that group A is larger than group B is 'one-tailed'. The effect of using a one-tailed test is to make statistics much less conservative for the same value of P

One-way ANOVA a parametric test of the null hypothesis that two or more groups come from the same population

Ordinal when values in a set of data can be placed in a meaningful order and ranked

Orthogonal literally means 'at right angles to'. In statistics it is used to indicate that two variables, factors or components are unrelated to each other

Outlier an extreme or aberrant observation lying well away from the rest of the data

P-value or p-value the probability of the significance statistic being that extreme or more if the null hypothesis is true. In biology the null hypothesis is usually rejected if the P-value is < 0.05

Paired observations two observations taken from the same individual, perhaps in a 'before and after' design (synonym for repeated measures)

Paired samples sets of data comprising paired observations

Paired t-test to test the null hypothesis that two sets of observations on the same individuals (e.g. before and after) have the same distribution

Parametric test one where assumptions about the shape and spread of the data are made within the test

Partial correlation a method of examining relationships between more than two variables that examines them pairwise while other variables are held constant

Path analysis a complex data exploration system where there are several simultaneous and possibly interacting 'cause' and 'effect' relationships

Pearson product–moment correlation coefficient the standard parametric correlation coefficient, r, measuring the association between two variables, that varies from 1 (perfect positive correlation), through 0 (no relationship), to -1 (perfect negative correlation)

Percentage when the relationship between two values is expressed as a single value (usually on a scale from 0 to 100)

Phi coefficient the result of a 2×2 contingency table

Pie chart a simple representation of frequencies of observations in different categories as sections (slices) of a circle. Particularly appropriate when the categories do not have a logical sequence

Pillai test or trace a method for calculating P-values in MANOVA

Platykurtic a distribution that has more observations in the 'shoulders' and fewer around the mean and in the tails than a normal distribution

Poisson distribution a discrete frequency distribution that results when events occur entirely at random

Polynomial regression a regression where the relationship between 'cause' and 'effect' is not assumed to be a straight line

Population the pool of possible individuals from which a sample is taken. Do not confuse with a biological population which will include additional individuals

Position the position of the sample is its mid-point which can be defined as a mean, median or mode

Post hoc **test** meaning 'after this'; there are several tests used after a one-way ANOVA to determine which groups are different from which. Common methods include the LSD, SNK and Bonferroni tests

Power analysis a method based on the number of observations for determining the likelihood of detecting a statistically significant effect

Precision the range of possible values between which a particular observation may lie; the repeatability of a measurement (different to accuracy which is a measure of the closeness of a measurement to the true value)

Prediction interval in regression the range of y values that are expected for a given value of x. Usually given as a 95% prediction interval. This range does not fall within a zone running parallel to the best-fit line in linear regression but is smaller around the mid-range of x

Predictor in regression the 'cause' variable is often called the predictor. Always plotted on the x-axis

Principal axes a synonym for principal components

Principal axis regression a model II regression technique

Principal component analysis a multivariate test which weights the variables to maximize the differences between individuals

Probit a transformation of data based on the use of cumulative probabilities and log graph paper

Procedural control a control where all the disturbance associated with applying a treatment is carried out, but the treatment itself is not applied

Proportion when the relationship between two values is expressed on a scale from 0 to 1 (e.g. 30 out of 60 becomes 0.5)

Proportional frequency a synonym for relative frequency

Pseudoreplication a problem in statistics when samples that are not independent are being treated as such

q–q plot a q–q or quantile–quantile plot is a graphical method for assessing whether a variable follows a normal distribution

Quadrat a square sampling area

Quadratic regression a regression where the relationship between the variables is assumed to be best described by a quadratic equation

Qualitative an observation that is assigned to a category that, although it may be coded as a number, has no numerical value (e.g. sex: coded 1 for female and 2 for male)

Quantitative an observation that has a meaningful numeric value. It can be either a direct observation or a count

Quartile when data are ranked the quartiles are the values of the data points 25% and 75% down the list. They form the limits for the interquartile range

r the result of a Pearson product–moment correlation. If $r = 0$ there is no correlation. If $r > 0$ there is a positive relationship and if $r < 0$ it is negative; 1 is perfect correlation, -1 is perfect negative correlation

r_s the statistic associated with the Spearman rank correlation test

R a shareware version of the statistical package S

Random effect a term applied to factors in ANOVA that are not set by the experimenter (in contrast to a fixed effect)

Random sample a sample where each individual in the population has an equal chance of being measured or collected.

Randomized block design when sampling units are placed into groups (blocks) and the treatment applied to each sample is randomized within the block

Range a crude measure of dispersion: the distance from the lowest to highest value in a data set

Ratio when two values are expressed as a single number (e.g. 6 : 2 becomes 3). Ratios lose information and magnify the error associated with measurement

Raw data observations as they were originally recorded before any transformations or other processing is applied

Reduced major axis regression a model II regression technique where the slope is essentially determined by the ratio of the standard deviations of the x and y values

Regression a description of the relationship between two variables where the value of one is determined by the value of the other (synonym of linear regression). More advanced regression can use several 'cause' variables

Related measures a synonym for paired samples or repeated measures

Related samples a synonym for paired samples or repeated measures

Relative frequency the proportion of observations having a particular value (or range of values). It is the frequency scaled to the sample size (e.g. 45 of 108 nests in a survey had four eggs, the relative frequency of four eggs is 45/108 or 0.417 or 41.7%)

Repeated measures two or more observations taken from the same individual, same site, same transect, etc. at different times (if only two observations then this is a synonym for paired samples)

Repeated measures ANOVA analysis of variance carried out using repeat observations of the same individual. Time of observation (e.g. before and after) will be used as one of the factors in the ANOVA but the degrees of freedom will be reduced

Residuals the variation in the data left over after a statistical model has been accounted for (often regression or ANOVA). The model with the best fit has the smallest residual variation

Response in regression the 'effect' variable is often called this. Always plotted on the *y*-axis

Ryan–Joiner test a method for determining whether a set of data follows a normal distribution

S a statistical package (also comes in S-plus version)

Sample as all the individuals in a population may rarely be counted a portion of the population has to be taken, this is a sample

Sample size the number of observations in a sample

Sample variance the variance of a single sample

Sampling unit the level at which an individual observation is made (e.g. a quadrat or a given size; a single leaf)

SAS a widely available and powerful statistical package

Scatter (plot) a graphical method for examining two (or possibly three) sets of data for possible relationships

Scheffé–Box test a test for heterogeneity of variance

Scheffé test one of many *post hoc* tests used in ANOVA to determine which groups are different from which

Scheirer–Ray–Hare test a non-parametric version of a two-way ANOVA, rarely supported in packages but quite easy to implement using the usual parametric ANOVA on ranked data and simple treatment of the resulting *F*-value

Second-order interaction an interaction between three factors in ANOVA

Sidák test a synonym for the Dunn–Sidák test for multiple comparisons

Sign test a very conservative non-parametric test of a null hypothesis that there is no difference between two groups

Significance level the probability of achieving a significant result if the null hypothesis is true. In biology this is usually set at 0.05

Significant when the null hypothesis is rejected because the *P*-value is less than 0.05 (the usual value in biology), 0.01, or any value set by the tester

Simple factorial (design) none of the grouping variables are subgroups of any others, i.e. there is no nesting

Single-classification ANOVA a synonym for one-way ANOVA (there is only one grouping variable)

Skew(ness) a measure of the symmetry of a data set (sometimes called g_1). Positive skew indicates that there are more values in the right tail of a distribution than would be expected in a normal distribution. Negative skew indicates more values in the left tail

Skewed distribution a distribution that has a value of skewness other than 0

Slope a number (usually b or β) denoting how a trend line deviates from zero (a slope of 0)

SNK test see Student–Newman–Keuls test

Spearman rank correlation an alternative name for the Spearman rank-order correlation

Spearman rank-order correlation a non-parametric measure of correlation

Split-plot design an experimental design technique used to analyse two factors when there is only one 'plot' for each level of one of the factors

Spread the way in which the data are distributed, often measured by standard deviation (synonym of dispersion)

SPSS a widely available statistical package

Standard deviation a measure of spread: sensitive to the shape of the distribution

Standard error a measure of spread: the standard deviation of the values of a set of means taken from a data set. Sensitive to sample size

Statistica a statistical package

Stem and leaf chart a method of displaying the data commonly used by computers before graphical output was possible. The 'stem' would be represented by a series of rows and 'leaf' by columns starting from the left. Each observation would be assigned to a position on the stem based on its value

Stepwise regression a regression analysis where the best method for predicting the 'effect' from several 'cause' variables is sought

Stratified random sample a method of collecting a sample that takes into account a feature of the collecting area

Student a pseudonym used by the statistician William Gossett

Student–Newman–Keuls (SNK) test a frequently used *post hoc* test, used after a one-way ANOVA to determine which groups are different from which

Student *t*-test a synonym for independent samples *t*-test

Summary statistic anything that condenses the information about a variable, such as a mean or standard deviation

Symmetry a data set with symmetry has the same shape either side of the mean

Systat a statistical package

***t*-distribution** a family of distributions widely used in statistics that is derived from the distribution of sample means with respect to the true mean of a population

***t*-test** to test the null hypothesis that two groups come from the same distribution (synonym for the Student *t*-test, independent samples *t*-test)

Tally when observations are assigned to categories and marked as ticks in a table

Three way . . . an experiment involving three independent factors; *see* entry for two way . . . and extrapolate

Time series a set of data points taken at different points in time

Transect a method of taking a sample of observations: usually by selecting a straight line between two random points, or from one random point in a random direction for a set distance

Transformation a mathematical conversion that is applied to every observation in a data set. Usually used to make a distribution conform to a normal distribution

Treatment a level (usually denoted by an integer) of an independent variable, factor or grouping variable (i.e. set or defined by the experimenter)

Tukey–Kramer method a synonym for the Tukey test. A *post hoc* test, used after a one-way ANOVA to determine which groups are different from which

Tukey test one of many *post hoc* tests used in ANOVA to determine which groups are different from which

Two-tailed test applies to most statistical tests and implies that the null hypothesis can be rejected by deviations either up or down. For example, if the null hypothesis is that two groups of bats use the same frequency for echolocation, and this is rejected then group 'A' may use either a significantly higher or significantly lower frequency than group 'B'. If the standard $P = 0.05$ level is used then it implies a $P = 0.025$ region in each tail

Two-way ANOVA an ANOVA test where there are two independent ways of grouping the data (two factors)

Two-way interaction in ANOVA this a measure of whether two grouping variables have an additive (no interaction) effect or not (interaction)

Type I error a truly non-significant result is deemed significant by a test

Type II error a truly significant result is deemed non-significant by a test

Unbalanced when there are different numbers of observations in different factor combinations. Severely unbalanced designs (i.e. where some of the factor combinations have no observations) should be avoided

Uniform distribution a 'flat' distribution where the chance of any value occurring is approximately equal, may often be transformed, using the arcsin transform, to an approximately normal distribution

Unimodal a frequency distribution with a single peak at the mode

Univariate statistics statistical tests using only one dependent variable

Unpaired data a synonym for independent data, stressing that sets of data are not paired

Value a single piece of data (datum)

Variable anything that varies between individuals (e.g. 'sex', 'weight' or 'aggressiveness'). The term variate is actually correct, but variable is now the widely used term for the observed data set

Variance the sum of squared deviations of observations from the mean: a measure of spread of the data. Very important in the mechanics of statistics but not very useful as a descriptive statistic

Variance/mean ratio (v/m or s^2/m) a commonly quoted descriptive statistic useful for determining whether a set of observations fits a Poisson distribution ($v/m = 1$), is more clumped ($v/m > 1$) or is more ordered ($v/m < 1$)

Variate the correct term for variable. Still retained for terms such as canonical variate analysis or univariate statistics

Weibull distribution a family of continuous distributions

Welch approximate *t*-test a version of the Student *t*-test that can be used when the variances of the two samples are known to be unequal

Welsch step-up procedure a *post hoc* test, used after a one-way ANOVA to determine which groups are different from which; requires equal sample sizes

Wilcoxon–Mann–Whitney test a synonym for the Mann–Whitney U test. A non-parametric test of a null hypothesis that two groups come from the same distribution

Wilcoxon rank sum *W* test a synonym for the Mann–Whitney *U* test. A non-parametric test of a null hypothesis that two groups come from the same distribution

Wilcoxon signed ranks test a non-parametric test of a null hypothesis that there is no difference between two related groups. The non-parametric equivalent of the paired *t*-test

Wilks test a method for calculating *P*-values in MANOVA

Williams correction a method of correcting bias in various contingency tests such as the G-test

Winsorize a method used to reduce the effect of outlying observations by replacing them with the next value towards the median

Within sometimes used as shorthand for 'within-sample variance' or 'within-group variance'

Within-sample variance in ANOVA a measurement of the amount of variation within a sample (*see* between-sample variance)

***x*-axis** the horizontal axis of a graph or chart (abscissa)

***y*-axis** the vertical axis of a graph or chart

Yates correction sometimes called the 'continuity correction'. A method to make the results of a 2 × 2 chi-square test more conservative

***z*-axis** the axis that 'goes into the paper or screen' on a 3D graph or chart

***z*-distribution** occasionally used as a synonym for the normal distribution

***z*-test** a test used to compare two distributions, or more usually to compare a sample with a larger population where the mean and standard deviation are known

12

Assumptions of the tests

Most statistical tests make assumptions about the data they are being applied to; if the assumptions are violated it is wise to treat the results with caution especially when *P*-values fall in the range 0.01 to 0.1.

Here is a test-by-test summary of the assumptions.

Test	Assumptions
G-test	Observations can be assigned to groups or categories
Chi-square test	Observations can be assigned to groups or categories
Kolmogorov–Smirnov	Observations come from a fairly continuous scale
paired *t*-test	Both sets of data are normally distributed; variance is the same in both samples (although there are often tests incorporated into statistical packages that make corrections)
Wilcoxon signed ranks test	Observations are made on a scale such that the magnitude of differences is meaningful
Sign test	Observations are made on a scale so that the question 'is A bigger than B?' can be answered
t-test	Both sets of data are normally distributed; variance is the same in both samples (although there are often tests incorporated into statistical packages that make corrections)
Mann–Whitney *U* test	Observations are made on a continuous scale (i.e. they can be put into rank order with very few ties)
Friedman test	One observation per factor combination; observations may be put in meaningful rank order
All ANOVA tests	Observations are independent both within and between samples; variance is the same in all samples; data are normally distributed; observations are assigned to groups (coded by integers) using one or more factors
Kruskal–Wallis test	Observations are made on a fairly continuous scale (i.e. they can be put into rank order with very few ties)
Scheirer–Ray–Hare test	Observations are made on a continuous scale (i.e. they can be put into rank order with very few ties)
Chi-square test of association	Observations can be assigned to categories or groups using one or more factors

Phi coefficient of association	Observations can be assigned to two groups for each of two factors
Cramér coefficient of association	Observations can be assigned to categories or groups using two factors
'Standard' correlation (Pearson product–moment correlation)	Individuals have observations for two variables measured on a continuous scale; the two variables are both normally distributed
Spearman rank-order correlation	Individuals have observations for two variables measured on an approximately continuous scale
Kendall rank-order correlation	Individuals have observations for two variables measured on an approximately continuous scale
Kendall robust line-fit method	'Effect' measured on an approximately continuous scale and 'cause' on any meaningful scale
ANCOVA (analysis of covariance)	Observations and covariate measured on a continuous scale; variance the same for all factor levels; observations normally distributed; observation independent
'Standard' regression (model I linear regression)	'Cause' (= independent or x) variable measured without error; variation in 'effect' (= dependent or y) the same for all values of 'cause'; relationship between x and y is linear; 'effect' is measured on a continuous scale; 'effect' should be normally distributed for any value of 'cause'
Logistic regression	'Cause(s)' (= independent or x) variable(s) measured without error, can be categorical variable(s); variation in 'effect' (= dependent or y) the same for all values of 'cause'; relationship between x and y is linear; 'effect' can be expressed as a proportion (and then transformed by logits), can be categorical variable
Model II regression	Individuals have observations for y variable measured on an approximately continuous scale
Polynomial regression	As standard regression but not assuming that the relationship between x and y is linear
Stepwise regression/ multiple regression/ path analysis	As standard regression but with several 'effect' variables measured for each individual
Discriminant function analysis	Individuals have two or more observations assigned to them measured on continuous scales
Principal component analysis or factor analysis	Individuals have two or more observations assigned to them measured on continuous scales
Canonical variate analysis	Individuals have two or more observations assigned to them; observations are measured on continuous scales; individuals can be assigned to groups

MANOVA (multivariate analysis of variance)	Two or more observations for each individual; observations are independent both within and between samples; observations are assigned to groups (coded by integers) using one or more factors; variance is the same in all samples; observations are normally distributed
MANCOVA (multivariate analysis of covariance)	Two or more observations for each individual; observations are independent both within and between samples; variance is the same in all samples; observations are normally distributed; observations are assigned to groups (coded by integers) using one or more factors; covariate is measured on a continuous scale
cluster analysis (a family of techniques that have slightly different assumptions)	Each individual has two or more observations assigned to it; observations are measured on meaningful scales; individuals can be assigned to groups

What if the assumptions are violated?

There are several possible courses of action that can be taken (in approximate order of preference):

1 Data could be transformed to make them suitable for the analysis chosen.

2 An alternative test of the same hypothesis but with different assumptions is used instead.

3 The hypothesis is reframed to allow a different test to be used.

4 Violation of the assumptions could be ignored totally but the results regarded with caution.

5 No test is carried out at all.

Hints and tips

Using a computer

• Save frequently — packages usually have a little disk icon on screen all the time, click on it! (Computers 'hang', 'lock' or 'crash' every now and again and you want to make sure you do not lose data.)

• Learn a few keyboard shortcuts.

• An easy way to select a block of text or data in many packages is to place the cursor at the beginning, move the pointer to the end and press shift as you left click the mouse.

• Another way to select blocks of text is to hold shift while moving the down arrow, up arrow, page up or page down.

• Using the underlines — the underlined letters in menus mean that you can access the menu by typing the letter on the keyboard while holding the 'alt' key.

• Use the 'tab' key to move between boxes — useful in many of the Windows dialogue boxes.

• Use 'shift' and 'tab' together to move backwards through boxes — useful to correct mistakes.

• Back up your important files frequently on floppy disks, CD, zip disk, etc.

• Keep back ups in a different place to avoid total loss from theft or fire.

• Holding 'alt' and pressing 'tab' moves you between open packages — this is extremely useful!

• Edit in the best editing package, then do the statistics or graph drawing in another — do not feel you have to use the pathetic spreadsheet capabilities of the statistics package.

• If you are given data in the format of another package that your package cannot read you can nearly always read it by saving in raw text format from the first package.

• When converting labels into numbers I suggest that using alphabetical order all the time will avoid many problems of converting the numbers back to labels.

• Cut and paste is a very powerful facility of most packages — you can usually copy material from one to another using copy and paste.

• The keyboard shortcuts for 'cut', 'copy' and 'paste' are nearly always 'ctrl + x', 'ctrl + c' and 'ctrl + v', respectively. Using the shortcuts is easier and quicker than going to the 'edit' menu and selecting from there.

- Double-clicking or right-clicking often brings up helpful options.
- In Excel, clicking the plain square on the top left of the spreadsheet between A and 1 selects all cells and allows you to change all fonts or column widths, etc.
- If you get stuck try the help file — these are usually extensive and often have examples too.
- 'alt + F4' will usually close a package.
- Do not leave lots of unnecessary windows open — they slow the computer down (e.g. graphs may be in separate windows).

Sampling

- Try to balance sampling designs if you can (i.e. take equal-sized samples).
- Measure everything you can easily — you never know what is going to be important.
- Avoid sampling at regular intervals if possible.
- Choosing the nearest individual to a random point will *always* bias the sample to individuals on the edge of clumps and against those in the middle.
- Do not carry out repeat sampling in the same sequence.
- Measure to sensible precision only, not to maximum.
- Check the quality of measurements by repeat measuring the same individual after an interval.
- Three subsamples from a site are much better than two.
- When setting up a laboratory experiment make sure it is genuinely factorial (i.e. there should be no confounding factors such as 'all species x were from the sunny site').
- Randomize measurement routines as much as possible (i.e. do not measure all of group 1 then 2, etc.).
- Try double-blind labelling if possible (i.e. when measuring you do not know what group the individual belongs to).
- Do not try to design overelaborate experiments — it is difficult to interpret anything with more than three factors.
- Use transects with caution as they can easily produce biased samples.
- If measurements are taken by several different people check the quality of the data by having everyone blind measure the same individuals.
- Always sample with a clear idea in mind of the statistical test you intend to use.

Statistics

- Try the analysis on dummy data before collecting any real observations.
- Find a design like one you have seen in a statistics book and try to repeat the result in your statistical package.

- Frame null hypotheses very carefully, before doing anything else.
- Always consider whether the data violate the assumptions of the test — if they do, be wary of the results.
- Transformation of the data can often turn an inappropriate data set into an appropriate one.
- One-tailed tests have their place (i.e. the alternative hypothesis is 'x is greater than y' rather than 'x is different to y') but if in any doubt use two-tailed tests.
- If P-values are close to 0.05 consider resampling to increase sample sizes.
- There is nothing 'special' about $P = 0.05$, so do not be completely tied to it.
- A P-value of 0.05 means a one in 20 chance of getting a result this, or more, significant even if the null hypothesis is true.
- In regression, if you are unsure which variable is the 'cause' and which is the 'effect' then the data are probably not suitable for regression anyway.
- If a non-parametric test is available use it.
- Carry out tests on incomplete data sets to get a feel for the results from the complete set.
- Use 95% confidence intervals when comparing several means rather than standard errors.
- The coefficient of variation is a good way to compare data sets with very different means.

Displaying the data

- Never use 3D effects for bar charts, pie charts, etc. (except, possibly, for posters).
- If you must use a three-axis graph make sure that every point is anchored to the 'floor' by a spike, otherwise there is no way of determining its position on *two* of the axes.
- Use the minimum amount of shading.
- Use black and white for everything (except, possibly, for posters).
- Avoid putting titles on graphs and figures.
- Use figure legends for every graph and make sure that the legend is informative enough to make the graph intelligible without reading the main text of a report.
- Do not use any more decimal places than you have to and, for raw data, no more than you have measured.
- If a graph has a measure of position (e.g. mean) then nearly always display a measure of dispersion as well.
- If you want the reader to compare figures make sure they have the same scales (especially on the y-axis).
- If you use a line graph it must be possible for intermediate values to exist as they are implied by the line!
- Do not be afraid to use log scales even when the observations are not logged, and remember that log 10 is easier for a reader to convert back to the original value.

- Never draw best-fit lines unless the data are suitable for regression.
- Never extend best-fit lines beyond the data.
- Always have gaps between bars on a bar chart (data are discrete).
- Never have gaps between bars in a histogram (data are continuous).
- Only use pie charts for categorical data that have no real scale.
- Do not clutter graphs with too much information.
- If plotting means always plot error bars too (95% confidence intervals or standard deviation).
- If plotting medians always plot quartiles too.

A table of
statistical tests

14

Tests of difference

Samples or groups	Factors	Data type	Fit to known distributions	One sample tests
1	–	Cat	G-test	fit to uniform: G-test, chi² test
		D	fit to Poisson: chi² test	e.g. median of 0?: Wilcoxon one sample test
		C	fit to normal: Kolmogorov–Smirnov test, Anderson–Darling test	e.g. mean of 0?: one sample t-test
			Unpaired data	**Paired data**
2	1	Cat	Chi² test	chi² test
		D	Mann–Whitney U test	Wilcoxon signed rank test
		C	t-test, one-way ANOVA	paired t-test
>2	1	Cat	Chi² test	chi² test
		D	Kruskal–Wallis test	Friedman test for repeated measures (no replication)
		C	one-way ANOVA	repeated measures ANOVA
2+	>1	D	two factors only: Friedman test if no replication: Scheirer–Ray–Hare	Friedman test for repeated measures (only one factor other than repeat and no replication)
		C	two-way ANOVA, or multiway ANOVA	repeated measures ANOVA

Analysis with covariate(s)

>1 group, 1+ factor, 1+ covariate	1 variable	ANCOVA
	>1 variable	MANCOVA

Tests of relationships

Correlation

Data type	Correlation	Samples or groups
Cat	Chi² for association	2
D	Kendall rank correlation, Spearman rank correlation	
C	Pearson product–moment correlation	

Regression

Data type	Regression	Samples or groups
D	logistic regression, model II regression, Kendall robust line fit	1 effect, 1 cause
C	linear regression, quadratic or polynomial regression	
C	multiple regression, stepwise regression	>1 effect, 1 cause

Multivariate tests

Data type	Multivariate tests	Samples or groups
many causes, many effects to explore	Path analysis	1+ group and any number of factors
many variables to explore	PCA	
groups to discriminate with many variables	CVA, discriminant function analysis, MANOVA, multiple regression, DCA	
groups to discriminate with discrete variables	TWINSPAN	
many proportion or categorical variables to explore	multiple logistic regression	

C, continuous data; Cat, categorical data; D, discrete data; although many tests will have an assumption that the data is normally distributed, if this is not the case then use the option indicated for discrete data.

Index

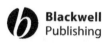